Entropy
Demystified

The Second Law Reduced to Plain Common Sense

Entropy Demystified

The Second Law Reduced to Plain Common Sense

Arieh Ben-Naim

The Hebrew University of Jerusalem, Israel

World Scientific

NEW JERSEY · LONDON · SINGAPORE · BEIJING · SHANGHAI · HONG KONG · TAIPEI · CHENNAI

Published by

World Scientific Publishing Co. Pte. Ltd.
5 Toh Tuck Link, Singapore 596224
USA office: 27 Warren Street, Suite 401-402, Hackensack, NJ 07601
UK office: 57 Shelton Street, Covent Garden, London WC2H 9HE

Library of Congress Cataloging-in-Publication Data
Ben-Naim, arieh, 1934–
    Entropy demystified : the second law reduced to plain common sense /
by Arieh Ben-Naim.
        p. cm.
    Includes bibliographical references.
    ISBN-13 978-981-270-052-0 (hardcover) -- ISBN-13 978-981-270-055-1 (pbk.)
    1. Entropy. 2. Second law of thermodynamics. I. Title.

    QC318.E57 B46 2007
    536'.73--dc22

                                                            2007011845

British Library Cataloguing-in-Publication Data
A catalogue record for this book is available from the British Library.

Typeset by Stallion Press
Email: enquiries@stallionpress.com

Printed by Mainland Press Pte Ltd

This book is dedicated to
Ludwig Boltzmann

Picture taken by the author, in Vienna, September 1978.

ORDERED

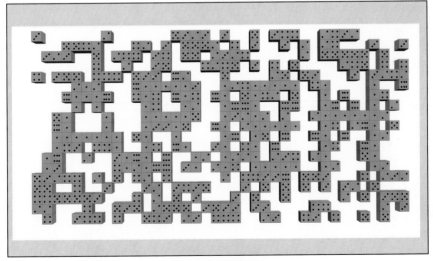

DISORDERED

Preface

Ever since I heard the word "entropy" for the first time, I was fascinated with its mysterious nature. I vividly recall my first encounter with entropy and with the Second Law of Thermodynamics. It was more than forty years ago. I remember the hall, the lecturer, even the place where I sat; in the first row, facing the podium where the lecturer stood.

The lecturer was explaining Carnot's cycle, the efficiency of heat engines, the various formulations of the Second Law and finally introducing the intriguing and mysterious quantity, named *Entropy*. I was puzzled and bewildered. Until that moment, the lecturer had been discussing concepts that were familiar to us; heat, work, energy and temperature. Suddenly, a completely new word, never heard before and carrying a completely new concept, was being introduced. I waited patiently to ask something, though I was not sure what the question would be. What is this thing called entropy and why does it always increase? Is it something we can see, touch or feel with any of our senses? Upon finishing her exposition, the lecturer interjected, "If you do not understand the Second Law, do not be discouraged. You are in good company. You will not be able to understand it at this stage, but you will understand it when you study statistical thermodynamics next year." With these concluding remarks, she had freed herself from any further explanation of the Second Law. The atmosphere was charged with mystery. I, as well as some of those present during the lecture were left tongue-tied, our intense craving for understanding the Second Law unsatisfied.

Years later, I realized that the lecturer was right in claiming that statistical mechanics harbors the clues to the understanding of entropy, and that without statistical mechanics, there is no way one can understand what lies beneath the concept of entropy and the Second Law. However, at that time, we all suspected that the lecturer had chosen an elegant way of avoiding any embarrassing questions she could not answer. We therefore accepted her advice, albeit grudgingly.

That year, we were trained to *calculate* the entropy changes in many processes, from ideal gas expansion, to mixing of gases, to transfer of heat from a hot to a cold body, and many other spontaneous processes. We honed our skills in *calculations* of entropy changes, but we did not really capture the essence of the meaning of entropy. We did the calculations with professional dexterity, pretending that entropy is just another technical quantity, but deep inside we felt that entropy was left ensconced in a thick air of mystery.

What is that thing called entropy? We knew it was *defined* in terms of heat transferred (reversibly) divided by the absolute temperature, but it was neither heat nor temperature. Why is it always increasing, what fuel does it use to propel itself upwards? We were used to conservation laws, laws that are conceived as more "natural." Matter or energy cannot be produced out of nothing but entropy seems to defy our common sense. How can a physical quantity inexorably keep "producing" more of itself without any apparent feeding source?

I recall hearing in one of the lectures in physical chemistry, that the entropy of solvation of argon in water is large and negative.[1] The reason given was that argon *increases* the *structure* of water. Increase of *structure* was tantamount to increase of order. Entropy was loosely associated with disorder. Hence, that was supposed to explain the *decrease* of entropy. In that class, our lecturer explained that entropy of a system *can* decrease when that system is coupled with another system (like a thermostat) and that the law of ever-increasing entropy is only valid in an isolated system — a system that does not interact with its surroundings. That fact only deepened the mystery. Not only do we not know the *source* which supplies the fuel for the ever-increasing entropy, but no source is permitted, in principle, no feeding mechanism and no provision for any supplies of anything from the outside. Besides, how is it that "structure" and "order" have crept into the discussion of entropy, a concept that was *defined* in terms of *heat* and *temperature*?

A year later, we were taught statistical mechanics and along side we learnt the relationship between entropy and the number of states, the famous Boltzmann relationship which is carved on Ludwig Boltzmann's tombstone in Vienna.[2] Boltzmann's relationship provided an interpretation of entropy in terms of disorder; the ever-increasing entropy, being interpreted as nature's way of proceeding from order to disorder. But why should a system go from order to disorder? Order and disorder are intangible concepts, whereas entropy was *defined* in terms of heat and temperature. The mystery of the perpetual increment of disorder in the system did not resolve the mystery of entropy.

I taught thermodynamics and statistical mechanics for many years. During those years, I came to realize that the mystery associated with the Second Law can never be removed within classical thermodynamics (better referred to as the

---

[1]This was another fascinating topic that was eventually chosen for my PhD thesis.
[2]A picture is shown on the dedication page of this book.

non-atomistic formulation of the Second Law; see Chapter 1).
On the other hand, looking at the Second Law from the molecular point of view, I realized that there was no mystery at all.

I believe that the turning point in my own understanding of
entropy, hence also in my ability to explain it to my students
came when I was writing an article on the entropy of mixing
and the entropy of assimilation. It was only then that I felt I
could penetrate the haze enveloping entropy and the Second
Law. It dawned on me (during writing that article) how two key
features of the atomic theory of matter were crucial in dispersing
the last remains of the clouds hovering above entropy; the large
(unimaginably large) numbers and the indistinguishability of the
particles constituting matter.

Once the haze dissipated, everything became crystal clear.
Not only clear, but in fact obvious; entropy's behavior which
was once quite difficult to understand, was reduced to a simple
matter of common sense.

Moreover, I suddenly realized that one *does not* need to
know any statistical mechanics to understand the Second Law.
This might sound contradictory, having just claimed that statistical mechanics harbors the clues to understanding the Second
Law. What I discovered was that, *all* one needs is the *atomistic formulation* of entropy, and nothing more from statistical mechanics. This finding formed a compelling motivation for
writing this book which is addressed to anyone who has never
heard of statistical mechanics.

While writing this book, I asked myself several times at
exactly what point in time I decided that this book was worth
writing. I think there were three such points.

First, was the recognition of the crucial and the indispensable
facts that matter is composed of a huge number of particles, and
that these particles are indistinguishable from each other. These
facts have been well-known and well-recognized for almost a

century, but it seems to me that they were not well emphasized by authors who wrote on the Second Law.

The second point was while I was reading the two books by Brian Greene.[3] In discussing the entropy and the Second Law, Greene wrote[4]:

> *"Among the features of common experience that have resisted complete explanation is one that taps into the deepest unresolved mysteries in modern physics."*

I could not believe that Greene, who has explained so brilliantly and in simple words so many difficult concepts in modern physics, could write these words.

The third point has more to do with aesthetics than substance. After all, I have been teaching statistical thermodynamics and the Second Law for many years, and even using dice games to illustrate what goes on in spontaneous processes. However, I always found the correspondence between the dice changing faces, and the particles rushing to occupy all the accessible space in an expansion process, logically and perhaps aesthetically unsatisfactory. As you shall see in Chapter 7, I made the correspondence between dice and particles, and between the outcomes of tossing dice and the *locations* of the particles. This correspondence is correct. You can always name a particle in a right compartment as an R–particle and a particle in the left compartment as an L–particle. However, it was only when I was writing the article on the entropy of mixing and entropy of assimilation, that I "discovered" a different process for which this correspondence could be made more "natural" and more satisfying. The process referred to is deassimilation. It is a spontaneous process where the change in entropy is due solely to

---

[3]Greene, B. (1999, 2004).
[4]Greene, B. (2004), p. 12.

the particles acquiring new identity. The correspondence was now between a die and a particle, and between the *identity* of the outcome of throwing a die, and the *identity* of the particle. I found this correspondence more aesthetically gratifying, thus making the correspondence between the dice-game and the real process of deassimilation a perfect one and worth publishing.

In this book, I have deliberately avoided a technical style of writing. Instead of teaching you what entropy is, how it changes, and most importantly why it changes in one direction, I will simply guide you so that you can *"discover"* the Second Law and obtain the satisfaction of unveiling the mystery surrounding entropy for yourself.

Most of the time, we shall be engaged in playing, or imagining playing, simple games with dice. Starting with one die, then two dice, then ten, a hundred or a thousand, you will be building up your skills in analyzing what goes on. You will find out what is that thing that changes with time (or with the number of steps in each game), and how and why it changes. By the time you get to a large number of dice, you will be able to extrapolate with ease whatever you have learned from a small number of dice, to a system of a huge number of dice.

After experiencing the workings of the Second Law in the dice world, and achieving full understanding of what goes on, there is one last step that I shall help you with in Chapter 7. There, we shall *translate* everything we have learned from the dice world into the real experimental world. Once you have grasped the evolution of the dice games, you will be able to understand the Second Law of thermodynamics.

I have written this book having in mind a reader who knows nothing of science and mathematics. The only prerequisite for reading this book is plain common sense, and a strong will to apply it.

One caveat before you go on reading the book; "common sense" does not mean easy or effortless reading!

There are two "skills" that you have to develop. The first is to train yourself to think in terms of big numbers, fantastically big numbers, inconceivably big numbers and beyond. I will help you with that in Chapter 2. The second is a little more subtle. You have to learn how to distinguish between a *specific* event (or state or configuration) and a *dim* event (or a state or configuration). Do not be intimidated by these technical sounding terms.[5] You will have ample examples to familiarize yourself with them. They are indispensable for understanding the Second Law. If you have any doubts about your ability to understand this book, I will suggest that you take a simple test.

Go directly to the end of Chapter 2 (Sections 2.7 and 2.8). There, you shall find two quizzes. They are specifically designed to test your understanding of the concepts of "specific" and "dim."

If you answer all the questions correctly, then I can assure you that you will understand the entire book easily.

If you cannot answer the questions, or if you tried but got wrong answers, do not be discouraged. Look at my answers to these questions. If you feel comfortable with my answers even though you could not answer the questions yourself, I believe you can read and understand the book, but you will need a little more effort.

If you do not know the answers to the questions, and even after reading my answers, you feel lost, I still do not think that understanding the book is beyond your capacity. I would suggest that you read Chapter 2 carefully and train yourself in thinking

---

[5] In statistical mechanics, these terms correspond to microstates and macrostates. In most of the book, we shall be playing with dice; and dice are always macroscopic. That is why I chose the terms "specific" and "dim" instead.

probabilistically. If you need more help, you are welcome to write to me and I promise to do my best to help.

Again, do not feel intimidated by the word "probabilistically." If you are not surprised that you did not win the one million prize in the lottery, although you habitually buy tickets, you have been thinking "probabilistically." Let me tell you a little story to make you comfortable with this formidable sounding word.

My father used to buy one lottery ticket every weekend for almost sixty years. He was sure that someone "up there" favored him and would bestow upon him the grand prize. I repeatedly tried to explain to him that his chances of winning the grand prize were very slim, in fact, less than one hundredth of one percent. But all my attempts to explain to him his odds fell on deaf ears. Sometimes he would get seven or eight matching numbers (out of ten; ten matches being the winning combination). He would scornfully criticize me for not being able to see the clear and unequivocal "signs" he was receiving from Him. He was sure he was on the right track to winning. From week to week, his hopes would wax and wane according to the number of matches he got, or better yet, according to the kind of signs he believed he was receiving from Him. Close to his demise, at the age of 96, he told me that he was very much disappointed and bitter as he felt betrayed and disfavored by the deity in whom he had believed all his life. I was saddened to realize that he did not, and perhaps could not, think *probabilistically*!

If you have never heard of the Second Law, or of entropy, you can read the brief, non-mathematical description of various formulations and manifestations of the Second Law in Chapter 1. In Chapter 2, I have presented some basic elements of probability and information theory that you might need in order to express your findings in probabilistic terms. You should realize that the fundamentals of both probability and information theory are based on nothing more than sheer common sense. You need not have any background in mathematics, physics or chemistry. The only things you need to know are: how to count (mathematics!), that matter is composed of atoms and molecules (physics and chemistry!), and that atoms are indistinguishable, (this is advanced physics!). All these are explained in non-mathematical terms in Chapter 2. From Chapters 3–5, we shall be playing games with a varying number of dice. You watch what goes on, and make your conclusions. We shall have plenty of occasions to "experience" the Second Law with all of our five senses. This reflects in a miniscule way the immense variety of manifestations of the Second Law in the real physical world. In Chapter 6, we shall summarize our findings. We shall do that in terms that will be easy to translate into the language of a real experiment. Chapter 7 is devoted to describing two simple experiments involving increase in entropy; all you have to do is to make the correspondence between the number of dice, and the number of particles in a box, between different outcomes of tossing a die, and the different states of the particles. Once you have made this correspondence, you can easily implement all that you have learned from the dice-game to understand the Second Law in the real world.

By the time you finish reading Chapter 7, you will understand what entropy is and how and why it behaves in an apparently capricious way. You will see that there is no mystery at all in its behavior; it simply follows the rules of common sense.

By understanding the two specific processes discussed in Chapter 7, you will clearly see how the Second Law works. Of course, there are many more processes that are "driven" by the Second Law. It is not always a simple, straightforward matter to show how the Second Law works in these processes. For this, you need to know some mathematics. There are many more, very complex processes where we *believe* that the Second

Law has its say, but there is, as yet, no mathematical proof of how it does that. Biological processes are far too complicated for a systematic molecular analysis. Although I am well aware that many authors do use the Second Law in conjunction with various aspects of life, I believe that at this stage, it is utterly premature. I fully agree with Morowitz[6] who wrote: "*The use of thermodynamics in biology has a long history of confusion.*"

In the last chapter, I have added some personal reflections and speculations. These are by no means universally accepted views and you are welcome to criticize whatever I say there. My email address is given below.

My overall objective in writing this book is to help you answer two questions that are associated with the Second Law. One is: *What* is entropy? The second is: *Why* does it change in only one direction — in apparent defiance of the time-symmetry of other laws of physics?

The second question is the more important one. It is the heart and core of the mystery associated with the Second Law. I hope to convince you that:

1. The Second Law is *basically* a law of probability.
2. The laws of probability are *basically* the laws of common sense.
3. It follows from (1) and (2) that the Second Law is *basically* a law of common sense — nothing more.

I admit, of course, that statements (1) and (2) have been stated many times by many authors. The first is implied in Boltzmann's formulation of the Second Law. The second has been expressed by Laplace, one of the founders of probability theory. Certainly, I cannot claim to be the first to make these statements. Perhaps I can claim that the relationship of

---

[6]Morowitz (1992) page 69.

"basicality" is a transitive relationship, i.e., that statement (3) follows from (1) and (2), is original.

The first question is about the *meaning* of entropy. For almost a hundred years, scientists speculated on this question. Entropy was interpreted as measuring disorder, mixed-upness, disorganization, chaos, uncertainty, ignorance, missing information and more. To the best of my knowledge, the debate is still on going. Even in recent books, important scientists express diametrically opposing views. In Chapter 8, I will spell out in details my views on this question. Here I will briefly comment that entropy can be made *identical*, both formally and conceptually, with a specific measure of information. This is a far from universally accepted view. The gist of the difficulty in accepting this identity is that entropy is a physically measurable quantity having units of energy divided by temperature and is therefore an *objective* quantity. Information however, is viewed as a nebulous dimensionless quantity expressing some kind of human attribute such as knowledge, ignorance or uncertainty, hence, a highly *subjective* quantity.[7]

In spite of the apparent irreconcilability between an objective and a subjective entity, I claim that entropy *is* information. Whether either one of these is objective or subjective is a question that encroaches on philosophy or metaphysics. My view is that both are objective quantities. But if you think one is subjective, you will have to concede that the second must be subjective too.

There is trade-off in order to achieve this identity. We need to redefine temperature in units of energy. This will require the sacrifice of the Boltzmann constant, which should have been *expunged* from the vocabulary of physics. It will bring a few other benefits to statistical mechanics. For the purpose of this

---

[7]More on this aspect of entropy may be found in Ben–Naim (2007).

book, absence of the Boltzmann constant will automatically make entropy dimensionless *and* identical with a measure information. This will, once and for all, "exorcise" the mystery out of entropy!

To the reader of this book, I dare to promise the following:

1. If you have ever learned about entropy and been mystified by it, I promise to unmystify you.
2. If you have never heard and never been mystified by entropy, I promise you immunity from any future mystification.
3. If you are somewhere in between the two, someone who has heard, but never learned, about entropy, if you heard people talking about the deep mystery surrounding entropy, then I promise you that by reading this book, you *should* be puzzled and mystified! Not by entropy, not by the Second Law, but by the whole ballyhoo about the "mystery" of entropy!
4. Finally, if you read this book carefully and diligently and do the small assignments scattered throughout the book, you will feel the joy of discovering and understanding something which has eluded understanding for many years. You should also feel a deep sense of satisfaction in understanding *"one of the deepest, unsolved mysteries in modern physics."*[8]

Acknowledgements

I want to express my sincerest thanks and appreciation to all those who were willing to read either parts, or the whole manuscript and offered comments and criticism.

I would like first of all to thank my friends and colleagues, Azriel Levy and Andres Santos for their meticulous reading and checking of the entire manuscript. They have saved me from

---

[8] Greene, B. (2004).

what could have been embarrassing errors I did not, or could not detect. Thanks are also due to Shalom Baer, Jacob Bekenstein, Art Henn, Jeffrey Gordon, Ken Harris, Marco Pretti, Samuel Sattath and Nico van der Vegt, who read parts of the manuscript and sent me valuable comments. Finally, I would like to express my sincere thanks to my lifetime partner Ruby for her patience in struggling with my messy handwriting and correcting and re-correcting the manuscript. Without her gracious help this manuscript could not have seen the light of publication. The book's planning had a long period of incubation. The actual writing of the book began in Burgos, Spain and it was completed in La Jolla, California, USA.

Arieh Ben-Naim
Department of Physical Chemistry
The Hebrew University
Jerusalem, Israel
Email: arieh@fh.huji.ac.il and ariehbennaim@yahoo.com

P.S.   Just in case you wonder about the meaning of the little figures at the end of each chapter, let me tell you that since I undertook the responsibility of explaining to you the Second Law, I decided to do a little espionage on your progress. I placed these icons so that I can monitor your progress in grasping the Second Law. You are welcome to compare your state of understanding with my assessment. If you do not agree, let me know and I will do my best to help.

Contents

Introduction, and a Short History of the Second Law of Thermodynamics

In this chapter, I shall present some important milestones in the history of the Second Law of Thermodynamics. I shall also present a few formulations of the Second Law in a descriptive manner. In doing so, I necessarily sacrifice precision. The important point here is not to teach you the Second Law, but to give you a qualitative description of the types of phenomena which led the scientists of the nineteenth century to formulate the Second Law.

There are many formulations of the Second Law of Thermodynamics. We shall group all these into two conceptually different classes: Non-Atomistic and Atomistic.

1.1. The *Non-Atomistic* Formulation of the Second Law[1]

Traditionally, the birth of the Second Law is associated with the name Sadi Carnot (1796–1832). Although Carnot himself did

---

[1]By "non-atomistic" formulation, I mean the discussion of the Second Law without any reference to the atomic constituency of matter. Sometimes, it is also said that this formulation views matter as a continuum. The important point to stress here is that these formulations use only macroscopically observable or measurable quantities without any reference to the atomic constituency of matter. It *does not* imply that the formulation applies to non-atomistic or continuous matter. As we shall see later, were matter really non-atomistic or continuous, the Second Law would not have existed.

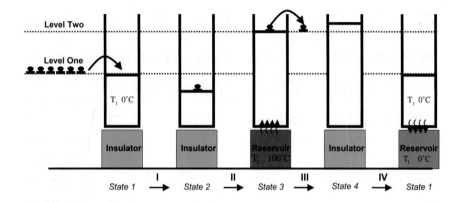

Fig. (1.1)   Heat engine.

not formulate the Second Law,[2] his work laid the foundations on which the Second Law was formulated a few years later by Clausius and Kelvin.

Carnot was interested in heat engines, more specifically, in the efficiency of heat engines. Let me describe the simplest of such an engine (Fig. (1.1)). Suppose you have a vessel of volume V containing any fluid, a gas or a liquid. The upper part of the vessel is sealed by a movable piston. This system is referred to as a heat engine. The vessel is initially in State 1, thermally insulated, and has a temperature T_1, say 0°C. In the first step of the operation of this engine (Step I), we place a weight on the piston. The gas will be compressed somewhat. The new state is State 2. Next, we attach the vessel to a heat reservoir (Step II). The heat reservoir is simply a very large body at a constant temperature, say $T_2 = 100$°C. When the vessel is attached to the heat reservoir, thermal energy will flow from the heat reservoir to the engine. For simplicity, we assume that the heat reservoir is immense compared with the size of the system or the engine. In Fig. (1.1), the heat reservoir is shown only at the bottom of the engine. Actually it should surround the entire engine. This

---

[2]This is the majority opinion. Some authors do refer to Carnot as the "inventor" or the "discoverer" of the Second Law.

ensures that after equilibrium is reached, the system will have the same temperature, T_2, as that of the reservoir, and though the reservoir has "lost" some energy, its temperature will be nearly unchanged. As the gas (or the liquid) in the engine heats up, it expands, thereby pushing the movable piston upwards. At this step, the engine did some useful work: lifting a weight placed on the piston from level one to a higher level, two. The new state is State 3. Up to this point, the engine has absorbed some quantity of energy in the form of heat that was transferred from the reservoir to the gas, thereby enabling the engine to do some work by lifting the weight (which in turn could rotate the wheels of a train, or produce electricity, etc.). Removing the weight, Step III, might cause a further expansion of the gas. The final state is State 4.

If we want to convert this device into an engine that repeatedly does useful work, like lifting weights (from level one to level two), we need to operate it in a complete cycle. To do this, we need to bring the system back to its initial state, i.e., cool the engine to its initial temperature T_1. This can be achieved by attaching the vessel to a heat reservoir or to a thermostat, at temperature $T_1 = 0°C$, Step IV (again, we assume that the heat reservoir is much larger compared with our system such that its temperature is nearly unaffected while it is attached to the engine). The engine will cool to its initial temperature T_1, and if we take away the weight, we shall return to the initial state and the cycle can start again.

This is not the so-called Carnot cycle. Nevertheless, it has all the elements of a heat engine, doing work by operating between the two temperatures, T_1 and T_2.

The net effect of the repeated cycles is that heat, or thermal energy, is pumped into the engine from a body at a high temperature $T_2 = 100°C$; work is done by lifting a weight and another amount of thermal energy is pumped out from the engine into a body at lower temperature $T_1 = 0°C$. The Carnot cycle is different in some details. The most important difference is that all the processes are done very gradually and very slowly.[3] We shall not be interested in these details here.

Carnot was interested in the *efficiency* of such an engine operating between two temperatures under some ideal conditions (e.g. mass-less piston, no friction, no heat loss, etc.).

At the time of the publication of Carnot's work in 1824,[4] it was believed that heat is a kind of fluid referred to as *caloric*. Carnot was mainly interested in the limits on the efficiency of heat engines. He found out that the limiting efficiency depends only on the ratio of the temperatures between which the engine operates, and not on the substance (i.e., which gas or liquid) that is used in the engine. Later, it was shown that the efficiency of Carnot's idealized engine could not be surpassed by any other engine. This laid the cornerstone for the formulation of the Second Law and paved the way for the appearance of the new term "entropy."

It was William Thomson (1824–1907), later known as Lord Kelvin, who first formulated the Second Law of Thermodynamics. Basically, Kelvin's formulation states that there could be no engine, which when operating in cycles, the *sole* effect of which is pumping energy from one reservoir of heat and completely converting it into work.

Although such an engine would not have contradicted the First Law of Thermodynamics (the law of conservation of the total energy), it did impose a limitation on the amount of work that can be done by operating an engine between two heat reservoirs at different temperatures.

---

[3]Technically, the processes are said to be carried out in a quasi-static manner. Sometimes, this is also referred to as a reversible process. The latter term is, however, also used for another type of process where entropy does not change. Therefore, the term quasi-static process is more appropriate and preferable.

[4]"Reflections on the motive power of fire and on machines fitted to develop this power," by Sadi Carnot (1824).

In simple terms, recognizing that heat is a form of energy, the Second Law of Thermodynamics is a statement that it is impossible to convert heat (thermal energy) completely into work (though the other way is possible, i.e., work can be converted completely into heat, for example, stirring of a fluid by a magnetic stirrer, or mechanically turning a wheel in a fluid). This impossibility is sometimes stated as "a perpetual motion of the second kind is impossible." If such a "perpetual motion" was possible, one could use the huge reservoir of thermal energy of the oceans to propel a ship, leaving a tail of slightly cooler water behind it. Unfortunately, this is impossible.

Another formulation of the Second Law of Thermodynamics was later given by Rudolf Clausius (1822–1888). Basically, Clausius' formulation is what every one of us has observed; heat always flows from a body at a high temperature (hence is cooled) to a body at a lower temperature (which is heated up). We never observe the reverse of this process occurring spontaneously. Clausius' formulation states that no process exists, such that its net effect is only the transfer of heat from a cold to a hot body. Of course we can achieve this direction of heat flow by doing work on the fluid (which is how refrigeration is achieved). What Clausius claimed was that the process of heat transferred from a hot to a cold body when brought in contact, which we observe to occur spontaneously, can never be observed in the reverse direction. This is shown schematically in Fig. (1.2), where two bodies initially isolated are brought into thermal contact.

While the two formulations of Kelvin and Clausius are different, they are in fact equivalent. This is not immediately apparent.

Fig. (1.2)

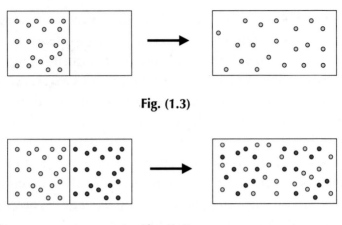

Fig. (1.3)

Fig. (1.4)

However, a simple argument can be employed to prove their equivalency, as any elementary textbook of thermodynamics will show.

There are many other formulations or manifestations of the Second Law of Thermodynamics. For instance, a gas in a confined volume V, if allowed to expand by removing the partition, will always proceed in one direction (Fig. (1.3)).[5] The gas will expand to fill the entire new volume, say $2V$. We never see a spontaneous reversal of this process, i.e., gas occupying volume $2V$ will never spontaneously converge to occupy a smaller volume, say V.

There are more processes which all of us are familiar with, which proceed in one way, never in the reverse direction, such as the processes depicted in Figs. (1.2), (1.3), (1.4) and (1.5). Heat flows from a high to a low temperature; material flows from a high to a low concentration; two gases mix spontaneously; and a small amount of colored ink dropped into a glass of water will spontaneously mix with the liquid until the water

---

[5]The Second Law may also be formulated in terms of the spontaneous expansion of a gas. It can also be shown that this, as well as other formulations, is equivalent to the Clausius and Kelvin formulations.

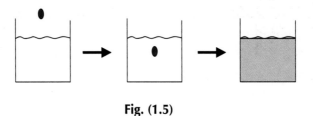

Fig. (1.5)

is homogeneously colored (Fig. (1.5)). We never see the reverse of these processes.

All these processes have one thing in common. They proceed in one direction, never proceeding *spontaneously* in the reverse direction. But it is far from clear that all these processes are driven by a common law of nature. It was Clausius who saw the general principle that is common in all these processes. Recall that Clausius' formulation of the Second Law is nothing but a statement of what everyone of us is familiar with. The greatness of Clausius' achievement was his outstanding prescience that all of these spontaneous processes are governed by one law, and that there is one quantity that governs the direction of the unfolding of events, a quantity that always changes in one direction in a spontaneous process. This was likened to a one-way arrow or a vector that is directed in one direction along the time axis. Clausius introduced the new term *entropy*. In choosing the word "entropy," Clausius wrote:[6]

> "*I prefer going to the ancient languages for the names of important scientific quantities, so that they mean the same thing in all living tongues. I propose, accordingly, to call S the entropy of a body, after the Greek word 'transformation.' I have designedly coined the word entropy to be similar to energy, for these two quantities are so*

---

[6]Quoted by Cooper (1968).

analogous in their physical significance, that an analogy of denominations seems to me helpful."

In the Merriam-Webster Collegiate Dictionary (2003), "entropy" is defined as: "change, literary turn, a measure of the unavailable energy in a closed thermodynamic system... a measure of the system's degree of order..."

As we shall be discussing in Chapter 8, the term *entropy* in the sense that was meant by Clausius is an inadequate term. However, at the time it was coined, the molecular meaning of entropy was not known nor understood. In fact, as we shall see later, "entropy" is not *the* "transformation" (nor the "change" nor the "turn"). It is something else that *transforms* or *changes* or *evolves* in time.

With the new concept of entropy one could proclaim the general overarching formulation of the Second Law. In any spontaneous process occurring in an isolated system, the entropy never decreases. This formulation, which is very general, embracing many processes, sowed the seed of the mystery associated with the concept of entropy, the mystery involving a quantity that does not subscribe to a conservation law.

We are used to conservation laws in physics. This makes sense:[7] material is not created from nothing, energy is not given to us free. We tend to conceive of a conservation law as "understandable" as something that "makes sense." But how can a quantity increase indefinitely and why? What fuels that unrelenting, ever-ascending climb? It is not surprising that the Second Law and entropy were shrouded in mystery. Indeed, within the context of the macroscopic theory of matter, the Second Law of Thermodynamics is unexplainable. It could have stayed

---

[7]Here we use the term "makes sense" in the sense that it is a common *experience* and not necessarily a consequence of logical reasoning.

a mystery forever had the atomic theory of matter not been discovered and gained the acceptance of the scientific community. Thus, with the macroscopic formulation we reach a dead end in our understanding of the Second Law of Thermodynamics.

1.2. The Atomistic Formulation of the Second Law

Before the development of the kinetic theory of heat (which relied on the recognition of the atomistic theory of matter), thermodynamics was applied without any reference to the composition of matter — as if matter were a continuum. Within this approach there was no further interpretation of entropy. That in itself is not unusual. Any law of physics reaches a dead end when we have to accept it as it is, without any further understanding. Furthermore, the Second Law was formulated as an absolute law — entropy *always* increases in a spontaneous process in an isolated system. This is not different from any other law, e.g. Newton's laws are *always* obeyed — no exceptions.[8]

A huge stride forward in our understanding of entropy and of the Second Law of Thermodynamics, was made possible following Boltzmann's statistical interpretation of entropy — the famous relationship between entropy and the total number of microstates of a system characterized macroscopically by a given energy, volume, and number of particles. Take a look at the cover illustration or at the picture of Boltzmann's statue. Ludwig Boltzmann (1844–1906),[9] along with Maxwell and many others, developed what is now known as the kinetic theory of gases, or the kinetic theory of heat. This not only led to the identification of temperature, which we can feel with

---

[8] "Always" in the realm of phenomena that were studied at that time, and which are now referred to as classical mechanics.

[9] For a fascinating story of Boltzmann's biography, see Broda (1983), Lindley (2001), and Cercignani (2003).

our sense of touch, with the motions of the particles constituting matter, but also to the interpretation of entropy in terms of the number of states that that are accessible to the system.

The atomistic formulation of entropy was introduced by Boltzmann in two stages. Boltzmann first defined a quantity he denoted as H, and showed that as a result of molecular collisions and a few other assumptions, this quantity always decreases and reaches a minimum at equilibirium. Boltzmann called his theorem "the minimum theorem", which later became famous as Boltzmann's *H-theorem* (published in 1872). Furthermore, Boltzmann showed that a system of particles starting with any distribution of molecular velocities will reach thermal equilibrium. At that point, H attains its minimum and the resulting velocity distribution will necessarily be the so-called Maxwell distribution of the velocities (see also Chapter 7).

At that time, the atomistic theory of matter had not yet been established nor universally accepted. Although the idea of the "atom" was in the minds of scientists for over two thousand years, there was no compelling evidence for its existence. Nevertheless, the kinetic theory of heat did explain the pressure and temperature of the gas. But what about entropy, the quantity that Clausius introduced without any reference to the molecular composition ofmatter?

Boltzmann noticed that his H-quantity behaved similarly to entropy. One needs only to redefine entropy simply as the negative value of H, to get a quantity that always *increases* with time, and that remains constant once the system reaches thermal equilibrium.

Boltzmann's *H-theorem* drew criticisms not only from people like Ernst Mach (1838–1916) and Wilhelm Ostwald

(1853–1932), who did not believe that atoms existed, but also from his colleagues and close friends.[10]

The gist of the criticisms (known as the reversibility objection or the reversibility paradox), is the seeming conflict between the so-called time-reversal[11] or time symmetry of the Newtonian's equations of motion, and the time asymmetry of the behavior of Boltzmann's H-quantity. This conflict between the reversibility of the molecular collisions, and the irreversibility of the H-quantity was a profound one, and could not be reconciled. How can one derive a quantity that distinguishes between the past and the future (i.e. always increasing with time), from equations of motions that are indifferent and do not care for the past and future? Newton's equations can be used to predict the evolution of the particles into the past as well as into the future. Woven into the H-*Theorem* were arguments from both mechanics and probability, one is deterministic and time symmetric, while the other is stochastic and time asymmetric. This conflict seems to consist of a fatal flaw in the Boltzmann H-*theorem*. It was suspected that either something was wrong with the H-*theorem*, or perhaps even with the very assumption of the atomistic nature of matter. This was clearly a setback for Boltzmann's H-*theorem* and perhaps a (temporary) victory for the non-atomists.

Boltzmann's reaction to the reversibility objection was that the H-*theorem* holds most of the time, but in very rare cases,

---

[10]For instance, Loschmidt wrote in 1876 that the Second Law cannot be a result of purely mechanical principle.

[11]It should be noted as Greene (2004) emphasized that "time-reversal symmetry" is not about time *itself* being reversed or "running" backwards. Instead, time reversal is concerned with whether events that happen *in* time in one particular temporal order can also happen in the reverse order. A more appropriate phrase might be "*event reversal* or *process reversal*".

it can go the other way, i.e. H might increase, or the entropy might decrease with time.

This was untenable. The (non-atomistic) Second Law of Thermodynamics, like any other laws of physics, was conceived and proclaimed as being absolute — no room for *exceptions*, not even rare exceptions. No one had ever observed violation of the Second Law. As there are no exceptions to Newton's equations of motion,[12] there should be no exceptions to the Second Law, not even in rare cases. The Second Law must be absolute and inviolable. At this stage, there were two seemingly different views of the Second Law. On the one hand, there was the classical, non-atomistic and absolute law as formulated by Clausius and Kelvin encapsulated in the statement that entropy never decreases in an isolated system. On the other hand, there was the atomistic formulation of Boltzmann which claimed that entropy increases "most of the time" but there are exceptions, albeit very rare exceptions. Boltzmann proclaimed that entropy could decrease — that it was not an *impossibility*, but only *improbable*.[13] However, since all observations seem to support the *absolute* nature of the Second Law, it looked as if Boltzmann suffered a defeat, and along with that, the atomistic view of matter.

In spite of this criticism, Boltzmann did not back down. He reformulated his views on entropy. Instead of the *H-theorem* which had one leg in the field of mechanics, and the other in the realm of probability, Boltzmann anchored both legs firmly on the grounds of probability. This was a radically

---

[12] Within classical mechanics.

[13] As we shall see in Chapters 7 and 8, the admitted non-absoluteness of the atomists' formulation of the Second Law is, in fact, more absolute than the proclaimed absoluteness of the non-atomists' formulation. On this matter, Poincare commented: "...to see heat pass from a cold body to a warm one, it will not be necessary to have the acute vision, the intelligence, and the dexterity of Maxwell's demon; it will suffice to have a little patience" quoted by Leff and Rex (1990).

new and foreign way of reasoning in physics. Probability, at that time, was not part of physics (it was not even a part of mathematics). Boltzmann proclaimed that entropy, or rather atomistic-entropy, is equal to the logarithm of the total number of arrangements of a system. In this bold new formulation, there were no traces of the equations of motion of the particles. It looks as if it is an ad-hoc new definition of a quantity, devoid of any physics at all, purely a matter of *counting* the number of possibilities, the number of states or the number of configurations. This atomistic entropy had built-in provisions for exceptions, allowing entropy to decrease, albeit with an extremely low probability. At that time, the exceptions allowed by Boltzmann's formulation seemed to *weaken* the validity of his formulation compared with the absolute and inviolable non-atomist formulation of the Second Law. In Chapter 8, I shall return to this point arguing that, in fact, the built-in provision for exceptions strengthens rather than weakens the atomistic formulation.

There seemed to be a state of stagnation as a result of the two irreconcilable views of the Second Law. It was not until the atomic theory of matter had gained full acceptance that the Boltzmann formulation won the upper hand. Unfortunately, this came only after Boltzmann's death in 1906.

A year earlier, a seminal theoretical paper published by Einstein on the Brownian motion provided the lead to the victory of the atomistic view of matter. At first sight, this theory seems to have nothing to do with the Second Law.

Brownian motion was observed by the English botanist Robert Brown (1773–1858). The phenomenon is very simple: tiny particles, such as pollen particles, are observed to move at seemingly random fashion when suspended in water. It was initially believed that this incessant motion was due to some tiny living organism, propelling themselves in the liquid. However,

Brown and others showed later that the same phenomenon occurs with inanimate, inorganic particles, sprinkled into a liquid.

Albert Einstein (1879–1955) was the first to propose a theory for this so-called Brownian motion.[14] Einstein believed in the atomic composition of matter and was also a staunch supporter of Boltzmann.[15] He maintained that if there are very large numbers of atoms or molecules jittering randomly in a liquid, there must also be fluctuations. When tiny particles are immersed in a liquid (tiny compared to macroscopic size, but still large enough compared to the molecular dimensions of the molecules comprising the liquid), they will be "bombarded" randomly by the molecules of the liquid. However, once in a while there will be assymetries in this bombardment of the suspended particles, as a result of which the tiny particles will be moving one way or the other in a zigzag manner.

In 1905 Einstein published as part of his doctoral dissertation, a theory of these random motions.[16] Once his theory was corroborated by experimentalists [notably by Jean Perrin (1870–1942)], the acceptance of the atomistic view became inevitable. Classical thermodynamics, based on the *continuous* nature of matter, does not have room for fluctuations. Indeed, fluctuations in a macroscopic system are extremely small. That is why we do not observe fluctuation in a macroscopic piece of matter. But with the tiny Brownian particles, the fluctuations

---

[14]It is interesting to note that the founders of the kinetic theory of gases such as Maxwell, Clausius and Boltzmann never published anything to explain the Brownian motion.

[15]It is interesting to note that Einstein, who lauded Boltzmann for his probabilistic view of entropy, could not accept the probabilistic interpretation of quantum mechanics.

[16]A well-narrated story of Einstein' theory of Brownian motion may be found in John Rigden (2005). A thorough and authoritative discussion of the theory of Brownian motion, including a historical background, has been published by Robert Mazo (2002).

are magnified and rendered observable. With the acceptance of the atomic composition of matter also came the acceptance of Boltzmann's expression for entropy. It should be noted that this formulation of entropy stood fast and was not affected or modified by the two great revolutions that took place in physics early in the 20th century: quantum mechanics and relativity.[17] The door to understanding entropy was now wide open.

The association of entropy with the number of configurations and probabilities was now unassailable from the point of view of the dynamics of the particles. Yet, it was not easily understood and accepted, especially at the time when probability was still not part of physics.

Almost at the same time that Boltzmann published his views on the Second Law, Willard Gibbs (1839–1903) developed the statistical mechanical theory of matter based on a purely statistical or probabilistic approach. The overwhelming success of Gibbs' approach, though based on probabilistic postulates,[18] has given us the assurance that a system of a very large number of particles, though ultimately governed by the laws of motion, will behave in a random and chaotic manner, and that the laws of probability will prevail.

The mere relationship between entropy and the number of states in a system is not enough to explain the behavior of

---

[17]Perhaps, it should be noted that within the recent theories of black holes, people speak about the "generalized Second Law of Thermodynamics" [Bekenstein (1980)]. It seems to me that this generalization does not affect Boltzmann's formula for the entropy.

[18]Today, any book on physics, in particular, statistical mechanics, takes for granted the atomic structure of matter. It is interesting to note in Fowler and Guggenheim's book on *Statistical Thermodynamics* (first published in 1939, and reprinted in 1956), one of the first *assumptions* is: "Assumption 1: The *atomistic constitution of matter*." They add the comment that "Today, this hardly ranks as an assumption but it is relevant to start by recalling that it is made, since any reference to atomic constitutions is foreign to classical thermodynamics." Today, no modern book on statistical mechanics makes that *assumption* explicitly. It is a universally accepted fact.

entropy. One must supplement this relationship with three critically important facts and assumptions. First, that there is a huge number of particles and an even "huger" number of microstates. Second, that all these states are equally likely i.e. have equal probability of occurrence, hence are equally likely to be visited by the system. Third, and most importantly, that at equilibrium, the number of *microstates* that are consistent with (or belonging to) the *macrostate* that we actually observe, is almost equal to the *total* number of possible microstates. We shall come back to these aspects of a physical system in Chapters 6 and 7.

With these further assumptions that would crystallize into a firm theory of statistical thermodynamics, the atomistic formulation of entropy has gained a decisive victory. The non-atomistic formulation of the Second Law is still being taught and applied successfully. There is nothing wrong with it except for the fact that it does not, and in principle cannot reveal the secrets ensconced in the concept of entropy.

Boltzmann's heuristic relation between entropy and the logarithm of the total number of states[19] did open the door to an understanding of the meaning of entropy. However, one needs to take further steps to penetrate the haze and dispel the mystery surrounding entropy.

There are several routes to achieve this end. I shall discuss the two main routes. One is based on the interpretation of entropy in terms of the extent of disorder in a system;[20] the second involves

---

[19]For simplicity and concreteness, think of N particles distributed in M cells. A full description of the state of the system is a detailed specification of which particle is in which cell.

[20]The association of entropy with disorder is probably due to Bridgman (1941;1953). Guggenheim (1949) suggested the term "spread" to describe the spread over a large number of possible quantum states. A thorough discussion of this aspect is given by Denbigh and Denbigh (1985).

the interpretation of entropy in terms of the missing information on the system.[21]

The first, the older and more popular route, has its origin in Boltzmann's own interpretation of entropy: a large number of states can be conceived of as having a large degree of disorder. This has led to the common statement of the Second Law of Thermodynamics that "Nature's way is to proceed from order to disorder."

In my opinion, although the order-disorder interpretation of entropy is valid in many examples, it is not always obvious. In a qualitative way, it can answer the question of *what* is the thing that changes in some spontaneous processes, but not in all. However, it does not offer any answer to the question of *why* entropy always increases.

The second route, though less popular among scientists is, in my opinion, the superior one. First, because *information* is a better, quantitative and objectively *defined* quantity, whereas order and disorder are less well-defined quantities. Second, information, or rather the missing information, can be used to answer the questions of *what* is the thing that changes in *any* spontaneous process. Information is a familiar word; like energy, force or work, it does not conjure up mystery. The measure of information is defined precisely within information theory. This quantity retains its basic meaning of *information* with which we are familiar in everyday usage. This is not the case when we use the concept of "disorder" to describe *what* is the thing that changes. We shall further discuss this aspect in Chapters 7 and 8. Information in itself does not provide an answer to the question of *why* entropy changes in this particular way. However, information unlike disorder, is defined in terms of probabilities

---

[21]Information theory was developed independently of thermodynamics by Claude Shannon in 1948. It was later realized that Shannon's informational measure is identical (up to a constant that determines the units) with Boltzmann's entropy.

and as we shall see, probabilities hold the clues to answering the question *"why."*

For these reasons, we shall devote the next chapter to familiarizing ourselves with some basic notions of probability and information. We shall do that in a very qualitative manner so that anyone with or without a scientific background can follow the arguments. All you need is sheer common sense. Once you acquire familiarity with these concepts, the mystery surrounding entropy and the Second Law will disappear, and you will be able to answer both the questions: "What is the thing that is changing?" and "Why is it changing in this particular manner?"

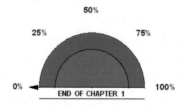

A Brief Introduction to Probability Theory, Information Theory, and all the Rest

Probability theory is a branch of mathematics. It has uses in all fields of science, from physics and chemistry, to biology and sociology, to economics and psychology; in short, everywhere and anytime in our lives.

We do probabilistic "calculations" or "assessments," consciously or unconsciously, in many decisions we make, whether it be crossing the street, taking a cab, eating never before tried food, going to war, negotiating a peace treaty and so on. In many activities we try to estimate the chances of success or failure.

Without this kind of probabilistic thinking, a doctor could not diagnose a disease from the symptoms, nor can he or she prescribe the best medication for a disease that has been diagnosed. Likewise, insurance companies cannot tailor-fit the cost of car insurance to different persons with varying personal profiles.

The theory of probability sprang from questions addressed to mathematicians by gamblers, presuming that the mathematicians have a better *knowledge* of how to *estimate* the chances of winning a game. Perhaps, some even believed that certain people have "divine" power and that they could *predict* the outcome of a game.[1]

Basically, probability is a subjective quantity measuring one's degree or extent of belief that a certain event will occur.[2] For instance, I may estimate that there is only a 10% chance that the suspect has committed a crime and therefore he or she should be acquitted. However, the judge may reach a completely different conclusion that the suspect, with high probability, was guilty. The reason that such an extreme discrepancy exists is mainly a result of different people having different information on the evidence and different assessment of this information. Even when two persons have the same information, they might process this information in such a way as to reach different estimates of the chances, or the probability of occurrence of an event (or the extent of plausibility of some proposition).

Out of this highly vague, qualitative and subjective notion, a distilled, refined theory of probability has evolved which is

---

[1]It is interesting to note that the Latin word for "guessing" is *adivinaré*, or in Spanish *adivinar*. The verb contains the root "divine." Today, when one says, "I guess," or when a Spanish speaking person says "*yo adivino*," it does not imply that one has some power to predict the outcome. Originally, the term *adivinaré* probably implied some divine power to predict the outcome of an experiment, or a game.

Bennett (1998) comments on this belief and writes: "*Ancients believed that the outcome of events was ultimately controlled by a deity, not by chance. The use of chance mechanism to solicit divine direction is called divination, and the step taken to ensure randomness were intended merely to eliminate the possibility of human interference, so that the will of the deity could be discerned.*"

[2]There is another, more general meaning of probability as a measure of the plausibility of a proposition, given some information or some evidence, Carnap (1950, 1953); Jaynes (2005). We shall use the term probability as used in physics. We shall always discuss events, not propositions. We shall not deal with the question of the meaning of probability, randomness, etc. These questions encroach upon philosophy. As we shall discuss in this chapter, questions about probability are always about conditional probability. Sometimes, the condition is formulated in terms of an event that has occurred or will occur. Other times, the condition can contain whatever information, or knowledge given on the event. Without *any* knowledge, no answer can be given to any question of probability.

quantitative and constitutes an *objective* branch[3] of mathematics. Although it is not applicable to all possible events, probability is applicable to a very large body of events; for instance, games of chance and many "events" which are the outcomes of experiments in physics.

Thus, if you claim that there is a probability of 90% that the Messiah will appear on Monday next week, and I claim that the chances are only 1%, there is no way to decide who is right or wrong. In fact, even if we wait for the coming Monday and see that nothing has happened, we could not tell whose estimate of the probability was correct.[4] However, for some classes of well-defined experiments, there *is* a probability that *"belongs"* to the event, and that probability is accepted by all.

For instance, if we toss a coin which we have no reason to suspect to be unbalanced or "unfair", the odds for the outcomes head (H) or tail (T) are 50%:50%, respectively. In essence, there is no proof that these are the "correct" probabilities. One can adopt a practical experimental proof based on actual, numerous

---

[3]It should be noted that "objective" here, does not imply "absolute probability." Any probability is a "conditional probability," i.e., given some information, or evidence. It is objective only in the sense that everyone would come to the same estimate of probability given the same information. D'Agostini (2003) has used the term "inter-subjectivity," others use the term "least subjective."

[4]There is a tendency, as one reader of this book commented, to conclude that the one who made the 1% guess was "correct," or "more correct" than the one who did the 90% guess. This is true if we use the term probability colloquially. However, here we are concerned with the scientific meaning of probability. Suppose I make a guess that the probability of obtaining the result "4" in throwing a die is 90%, and you make the guess that the probability is 1%. We throw the die and the result is "4" (or 2, or any other result). Whose guess was correct? The answer is neither! In this case, we know the probability of that particular event and the fact that the event "outcome 4" occurred in a single throw does not prove anything regarding the probability of that event. In the question we posed about the Messiah, the occurrence or the non-occurrence of the event does not tell us anything about the *probability* of that event. In fact, it is not clear how to define the probability of that event or even if there exists a "correct" probability of that event.

tossing of a coin and counting of the frequencies of the outcomes. If we toss a coin a thousand times, there is a good chance that about 500 outcomes will turn out to be H and about 500 will turn out to be T; but there is also a chance that we will get 590 Hs and 410 Ts. In fact, we can get any sequence of Hs and Ts by tossing the coin a thousand times; there is no way to *derive* or to extract the probabilities from such experiments. We must accept the existence of the probabilities in this, and similar experiments with dice, axiomatically. The odds of 50:50 per cent, or half for H and half for T, must be accepted as something belonging to the event, much as a quantity of mass belongs to a piece of matter. Today, the concept of probability is considered to be a primitive concept that cannot be defined in terms of more primitive concepts.

Let us go back to the pre-probability theory era from the 16th and 17th centuries, when the concept of probabilities was just beginning to emerge.

An example of a question allegedly addressed to Galileo Galilei (1564–1642) was the following:

Suppose we play with three dice and we are asked to bet on the *sum* of the outcomes of tossing the three dice simultaneously. Clearly, we feel that it would not be wise to bet our chances on the outcome of 3, nor on 18; our feeling is correct (in a sense discussed below). The reason is that both 3 and 18 have only one way of occurring; 1:1:1 or 6:6:6, respectively, and we intuitively judge that these events are relatively rare. Clearly, choosing the sum, 7, is better. Why? Because there are more *partitions* of the number 7 into three numbers (between 1 and 6), i.e., 7 can be obtained as a result of four possible partitions: 1:1:5, 1:2:4, 1:3:3, 2:2:3. We also *feel* that the larger the sum, the larger the number of partitions, up to a point, roughly at the center between the minimum of 3 and the maximum of 18. But how can we choose between 9 and 10? A simple count shows that both 9 and 10 have the same number of partitions, i.e., the same number of combinations of integers (from 1 to 6), the sum of which is 9 or 10. Here are all the possible partitions:

For 9: 1:2:6, 1:3:5, 1:4:4, 2:2:5, 2:3:4, 3:3:3

For 10: 1:3:6, 1:4:5, 2:2:6, 2:3:5, 2:4:4, 3:3:4

At first glance, we might conclude that since 9 and 10 have the same number of partitions, they should also have the same chances of winning the game. This conclusion would be wrong as discussed below. The correct answer is that 10 has better chances of winning than 9. The reason is that, though the number of partitions is the same for 9 and 10, the total number of outcomes of the three dice that sum up to 9, is a little bit smaller than the number of outcomes for 10. In other words, the number of partitions is the same, but each partition has a different "weight," e.g., the outcome 1:4:4 can be realized in three different ways:

$$1:4:4, \quad 4:1:4, \quad 4:4:1$$

This is easily understood if we use three dice having different colors, say blue, red and white, the three possibilities for 1:4:4 are:

blue 1, red 4 and white 4

blue 4, red 1 and white 4

blue 4, red 4 and white 1

When we count all the possible partitions and all the possible weights, we get the results shown below.

All the possible outcomes for *sum* = 9, for three dice:

1:2:6,	1:3:5,	1:4:4,	2:2:5,	2:3:4,	3:3:3
1:6:2,	1:5:3,	4:1:4,	2:5:2,	2:4:3	
2:1:6,	3:1:5,	4:4:1,	5:2:2,	3:2:4	
2:6:1,	3:5:1,			3:4:2	
6:1:2,	5:1:3,			4:2:3	
6:2:1,	5:3:1			4:3:2	

Weights:	6	6	3	3	6	1

Total number of outcomes for 9 is 25.

All the possible outcomes for *sum* = 10, for three dice:

1:3:6,	1:4:5,	2:2:6,	2:3:5,	2:4:4,	3:3:4
1:6:3,	1:5:4,	2:6:2,	2:5:3,	4:2:4,	3:4:3
3:1:6,	4:1:5,	6:2:2,	3:2:5,	4:4:2,	4:3:3
3:6:1,	4:5:1,		3:5:2,		
6:1:3,	5:1:4		5:2:3		
6:3:1,	5:4:1		5:3:2		

Weights:	6	6	3	6	3	3

Total number of outcomes for 10 is 27.

The total distinguishable outcome for the sum of 9 is 25, and for the sum of 10 is 27. Therefore, the relative chances of winning with 9 and 10, is 25:27, i.e., favoring the choice of 10. Thus, the best choice of a winning number, presumably as suggested by Galilei, is 10.

But what does it mean that 10 is the "best" choice and that this is the "correct" winning number? Clearly, I could choose 10 and you could choose 3 and you might win the game. Does our calculation guarantee that if I choose 10, I will always win? Obviously not. So what does the ratio 25:27 mean?

The theory of probability gives us an answer. It is not a precise nor a fully satisfactory answer, and it does not guarantee winning; it only says that if we play this game many times, the probability that the choice of 9 will win is $25/216$, whereas the probability that the choice of 10 will win is slightly larger, $27/216$ [216 being the total number of possible outcomes; $6^3 = 216$]. How many times do we have to play in order to guarantee my winning? On this question, the theory is mute. It only says that in the limit of an infinite number of games, the frequency of occurrence of 9 should be $25/216$, and the frequency of occurrence of 10 should be $27/216$. But an infinite number of games cannot be realized. So what is the meaning of these probabilities? At this moment, we can say nothing more than that the ratio 27:25, reflects our *belief* or our degree of confidence that the number 10 is more likely to win than the number 9.[5]

We shall leave this particular game for now. We shall come back to this and similar games with more dice later on.

In the aforementioned discussion, we have used the term probability without defining it. In fact, there have been several attempts to *define* the term probability. As it turns out, each definition has its limitations. But more importantly, each definition uses the concept of probability in the very definition, i.e., all definitions are circular. Nowadays, the mathematical theory of probability is founded on an axiomatic basis, much as Euclidian geometry or any other branch of mathematics is founded on axioms.

The axiomatic approach is very simple and requires no knowledge of mathematics. The axiomatic approach was

---

[5]Note that this statement sounds highly subjective. However, this subjectivity should be accepted by anyone who has common sense and who wants to use the theory of probability.

developed mainly by Kolmogorov in the 1930s. It consists of the following three basic concepts:

1) *The sample space*. This is the set of all possible outcomes of a specific, well-defined experiment. Examples: The sample space of throwing a die consists of six possible outcomes {1, 2, 3, 4, 5, 6}; tossing a coin has the sample space consisting of two outcomes {H:T} (H for head and T for tail). These are called *elementary events*. Clearly, we cannot write down the sample space for every experiment. Some consist of an infinite number of elements (e.g., shooting an arrow at a circular board); some cannot even be described. We are interested only in simple spaces where the counting of the outcomes, referred to as elementary events, is straightforward.

2) *A collection of events*. An *event* is defined as a union, or a sum of elementary events. Examples:

(a) The result of tossing a die is "even." This consists of the elementary events {2, 4, 6}, i.e., either 2 or 4 or 6 has occurred, or will occur in the experiment of tossing a die.[6]

(b) The result of tossing a die is "larger than or equal to 5." This consists of the elementary events {5, 6}, i.e., either 5 or 6 has occurred.

In mathematical terms, the collection of events consists of all partial sets of the sample space.[7]

3) *Probability*. To each event, we *assign* a number, referred to as the probability of that event, which has the following properties:

(a) The probability of each event is a number between zero and one.

----

[6]The use of the past or future does not imply that time is part of the definition of an event or of probability.

[7]We shall use only finite sample spaces. The theory of probability also deals with infinite, or continuous spaces, but we shall not need these in this book.

(b) The probability of the *certain* event (i.e., that any of the outcomes has occurred) is one.

(c) The probability of the *impossible* event is zero.

(d) If two events are disjoint or mutually exclusive, then the probability of the sum (or union) of the two events is simply the sum of the probabilities of the two events.

Condition (a) simply gives the scale of the probability function. In daily life, we might use the range 0–100% to describe the chances of, for example, raining tomorrow. In the theory of probability, the range (0,1) is used. The second condition simply states that if we *do* perform an experiment, one of the outcomes must occur. This event is called the certain event and is assigned the number one. Similarly, we assign the number zero to the impossible event. The last condition is intuitively self-evident. Mutual exclusivity means that the occurrence of one event excludes the possibility of the occurrence of the second. In mathematical terms, we say that the *intersection* of the two events is empty (i.e., contains no elementary event).

For example, the two events:

$$A = \{\text{the outcome of throwing a die is } even\}$$
$$B = \{\text{the outcome of throwing a die is } odd\}$$

Clearly, the events A and B are disjoint; the occurrence of one *excludes* the occurrence of the other. If we define the event:

$$C = \{\text{the outcome of throwing a die is}$$
$$\text{larger than or equal to } 5\}$$

Clearly, A and C, or B and C are not disjoint. A and C contain the elementary event 6. B and C contain the elementary event 5.

The events, "greater than or equal to 4," and "smaller than or equal to 2," are clearly disjoint. In anticipating the discussion below, we can calculate the probability of the first event $\{4, 5, 6\}$ to be $3/6$, and the probability of the second event $\{1, 2\}$ to be $2/6$; hence, the combined (or the union) event $\{1, 2, 4, 5, 6\}$ has the probability $5/6$, which is the sum of $2/6$ and $3/6$.

A very useful way of demonstrating the concept of probability and the sum rule is the Venn diagram. Suppose blindfolded, we throw a dart at a rectangular board having a total area of $S = A \times B$. We assume that the dart *must* hit some point within the board (Fig. (2.1)). We now draw a circle within the board,

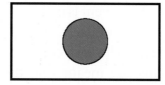

Fig. (2.1)

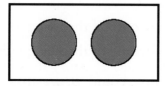

Fig. (2.2)

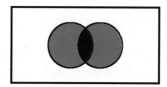

Fig. (2.3)

and ask what the probability of hitting the area within this circle is.[8] We assume by plain common sense, that the probability of the event "hitting inside the circle" is equal to the ratio of the area of the circle to the area of the entire board.[9]

Two regions drawn on the board are said to be disjoint if there is no overlap between the regions (Fig. (2.2)). It is clear that the probability of hitting either one region or the other is the ratio of the area of the two regions, to the area of the whole board.

This leads directly to the sum rules stated in the axioms above. The probability of hitting either one of the regions is the sum of the probabilities of hitting each of the regions.

This sum rule does not hold when the two regions overlap, i.e., when there are points on the board that belong to both regions, like the case shown in Fig. (2.3).

It is clear that the probability of hitting either of the regions is, in this case, the sum of the probabilities of hitting each of the regions, minus the probability of hitting the overlapping region. Simply think of the area covered by the two regions; it is the sum of the two areas of the two regions — minus the area of the intersection.

On this relatively simple axiomatic foundation, the whole edifice of the mathematical theory of probability has been erected. It is not only extremely useful but also an essential tool in all the sciences and beyond. As you must have realized, the

---

[8]We exclude from the present discussion the question of hitting exactly a specific point, or exactly a specific line, like the perimeter of the circle, the probability of which is negligible in practice, and zero in theory. We shall use the Venn diagrams only for illustrations. In actual calculations of probabilities in the forthcoming chapters, we shall always use finite sample spaces.

[9]Actually, we are asking about the conditional probability that the dart hit the circle, given that the dart has hit the board. See below on conditional probability.

basics of the theory are simple, intuitive, and require no more than common sense.

In the axiomatic structure of the theory of probability, the probabilities are said to be *assigned* to each event.[10] These probabilities must subscribe to the four conditions a, b, c, and d. The theory does not *define* probability, nor does it provide a method for calculating or measuring probabilities.[11] In fact, there is no way of calculating probabilities for any general event. It is still a quantity that measures our degree or extent of belief of the occurrence of certain events, and as such, it is a highly subjective quantity. However, for some simple experiments, say tossing a coin, throwing a die, or finding the number of atoms in a certain region of space, we have some very useful methods of calculating the probabilities. They have their limitations and they apply to "ideal" cases, yet these probabilities turn out to be extremely useful. What is more important, since these are based on common sense reasoning, we should *all* agree that these are the "correct" probabilities, i.e., these probabilities turn from being subjective quantities to objective quantities. We shall describe two very useful "definitions" that have been suggested for this concept.

---

[10]In mathematical terms, probability *is* a *measure* defined for each event. Much as the length, the area or the volume of a region in one, two or three dimensions, respectively. In the example using the Venn diagrams, we also take the *area* of a region as a measure of the relative probability.

[11]In fact, even Kolmogorov himself was well aware that he left the question of the meaning or the definition of probability unanswered! It is now almost universally accepted that probability is an un-definable primitive. Some authors of textbooks on probability even retrained from defining the term probability.

2.1. The Classical Definition

This definition is sometimes referred to as the *a priori* definition.[12] Let $N(total)$ be the *total* number of possible outcomes of a specific experiment. For example, $N(total)$ for throwing a die is six, i.e., the six outcomes (or six elementary events) of this experiment. We denote by $N(event)$, the number of outcomes (i.e., elementary events) that are included in the event that we are interested in. For example, the number of elementary events included in the event "even" is 3, i.e., $\{2, 4, 6\}$. The probability of the "event," in which we are interested, is *defined* as the ratio $N(event)/N(total)$. We have actually used this intuitively appealing definition when calculating the probability of the event "greater than or equal to 4." The total number of elementary events (outcomes) of throwing a die is $N(total) = 6$. The number of elementary events included in the event "greater than or equal to 4" is $N(event) = 3$, hence, the probability of this event is $3/6$ or $1/2$, which we all agree is the "correct" probability.

However, care must be taken in applying this *definition* of probability. First, not every event can be "decomposed" into elementary events, e.g. the event "tomorrow, it will start raining at 10 o'clock." But more importantly, the above formula presumes that each of the elementary events has the same likelihood of occurrence. In other words, each elementary event is presumed to have the same *probability*; $1/6$ in the case of a die. But how do we know that? We have given a formula for

---

[12]Some authors object to the use of the term "*a priori*." Here, we use that term only in the sense that it does not rely on an experiment to find out the probabilities. Also, the term "classical" is not favored by some authors. D'Agostini (2003) prefers to call this method the "combinatorial" method. It should be noted, however, that "combinatorics" is an exact branch of mathematics dealing with the number of ways of doing certain things. As such, it has nothing to do with probability. However, in probability, one uses the *combinatorial* method to calculate probabilities according to the *classical* definition.

calculating the probability of an event based on the knowledge of the probabilities of each of the elementary events. This is the reason why the classical definition cannot be used as a bona fide *definition* of probability; it is a circular definition. In spite of that, this "definition" (or rather the method of calculating probabilities) is extremely useful. Clearly, it is based on our *belief* that each elementary event has an equal probability, $1/6$. Why do we believe in that assertion? The best we can do is to invoke the argument of symmetry. Since all faces are presumed equivalent, their probabilities must be equal. This conclusion should be universally agreed upon, as much as the axiomatic assertion that two straight lines will intersect at most, at a single point. Thus, while the probability of the event "it will rain tomorrow" is highly subjective, the probability that the outcome of the event "even" in throwing a die is $1/2$, should be agreed upon by anyone who intends to use the probabilistic reasoning, as much as anyone who intends to use geometrical reasoning should adopt the axioms of geometry.

As in geometry, all of the probabilities as well as all the theorems derived from the axioms apply strictly to ideal cases; a "fair" die, or a "fair" coin. There is no definition of what a fair die is. It is as much an "ideal" concept as an ideal or Platonic circle or cube.[13] All *real* dice, as all cubes or spheres, are only approximate replicas of the ideal Platonic objects. In practice, if we do not have any reason to suspect that a die is not homogenous or unsymmetrical, we can assume that it is ideal.

In spite of this limitation, this procedure of calculating probabilities is very useful in many applications. One of the basic

---

[13] Of course, one assumes not only that the die is fair, but also that the method of throwing the die is "fair" or unbiased. The definition of a "fair" die, or the "random" toss of it, also involves the concept of probability. We might also add that information theory provides a kind of "justification" for the choice of equal probabilities. Information theory provides a method of guessing the best probabilities based on what we know, all we know, and nothing but what we know on the experiment.

postulates of statistical mechanics is that each of the microstates comprising a macroscopic system has the same probability. Again, one cannot prove that postulate much less than one can "prove" the assertion that each outcome of throwing a die is $1/6$. This brings us to the second "definition," or if you like, the second procedure of calculating probabilities.

2.2. The Relative Frequency Definition

This definition is referred to as the *a posteriori* or "experimental" definition since it is based on actual counting of the relative frequency of the occurrence of events.

The simplest example would be tossing a coin. There are two possible outcomes; head (H) or tail (T). We exclude the rare events, such as the coin falling exactly perpendicular to the floor, breaking into pieces during the experiment, or even disappearing from sight so that the outcome is indeterminable.

We proceed to toss a coin N times. The frequency of occurrence of heads is recorded. This is a well-defined and feasible experiment. If $n(H)$ is the number of heads that occurred in $N(total)$ trials, then the frequency of occurrence of head is $n(H)/N(total)$. The *probability* of occurrence of the event "H," is *defined* as the limit of this frequency when N tends to infinity.[14] Clearly, such a definition is not practical; first, because

---

[14]The frequency definition is

$$\Pr(H) = \operatorname*{limit\ of}_{N(total)\to\infty} \frac{n(H)}{N(total)}$$

This may be interpreted in two different ways. Either one performs a sequence of experiments, and measure the limit of the relative frequency when the number of experiments is infinity, or throw infinite coins at once and count the fraction of coins which turned out to be H. One of the fundamental assumptions of statistical mechanics is that average quantities calculated by either methods will give the same result. This hypothesis is the seed of a whole branch of mathematics known as Ergodic theory.

we cannot perform infinite trials. Second, even if we could, who could guarantee that such a limit exists at all? Hence, we can only imagine what this limit will be. We believe that such a limit exists, and it is unique; but, in fact, we cannot prove that.

In practice, we do use this definition for a very large number N. Why? Because we believe that if N is large enough and if the coin is fair, then there is a *high probability* that the relative frequency of occurrence of heads will be $1/2$.[15] We see that we have used the *concept* of probability again in the very definition of probability.

This method could be used to "prove" that the probability of each outcome of throwing a die is $1/6$. Simply repeat the experiment many times and count the number of times that the outcome 4 (or any other outcome) has occurred. The relative frequency can serve as "proof" of the probability of that event. This reasoning rests on the *belief* that if N is large enough, we should get the frequency of one out of six experiments. But what if we do the experiment a million times and find that the result "4" occurred in 0.1665 of the times (instead of 0.1666...)? What could be concluded? One conclusion could be that the die is "fair," but we did not run enough experiments. The second conclusion could be that the die is unfair, and that it is slightly heavier on one side. The third conclusion could be that the throwing of the die was not perfectly random. So, how do we estimate the probability of an event? The only answer we can give is that we believe in our *common sense*. We use common sense to judge that because of the symmetry of the die (i.e., all faces are equivalent), there must be equal probability for the outcome of any specific face. There is no way we can *prove* that.

---

[15]In fact, we believe that this is the right result even without actually doing the experiment. We are convinced that by doing a mental experiment, the result will converge to the right probability. If we do the experiment of say, throwing a die, and find that the frequency of occurrence of the event "even" is indeed nearly $1/2$, which is consistent with the result we have calculated from the classical definition, then we get further support to the "correctness" of our assignment of probabilities.

All we can say is that if the die is ideal (such a die does not exist), then we *believe* that if we toss the die many times, the outcome of say, 4, showing up in the long run is $1/6$ of the total number of experiments. This belief, though it sounds subjective, must be shared by all of us and regarded as *objective*. You have the right not to agree to this. However, if you do not agree to this, you cannot use the theory of probability, nor be convinced by the arguments presented in the rest of the book.

It should be noted, however, that the identification of the elementary events is not always simple or possible. We present one famous example to demonstrate this. Suppose we have N particles (say electrons) and M boxes (say energy levels). There are different ways of distributing the N particles in the M boxes. If we have no other information, we might assume that all the possible configurations have equal likelihood. Figure (2.4) shows all the possible configurations for $N = 2$ particles in $M = 4$ boxes.

We can assign equal probabilities to all of these 16 configurations. This is referred to as "classical statistics" — not to be confused with the "classical" definition of probability. This would be true for coins or dice distributed in boxes. It would not work for molecular particles distributed in energy levels.

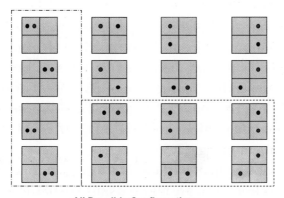

All Possible Configurations.

Fig. (2.4)

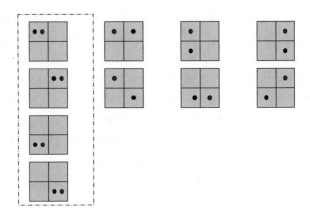

Bose Einstein Configurations.

Fig. (2.5)

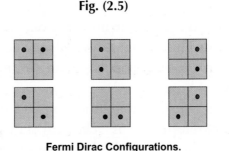

Fermi Dirac Configurations.

Fig. (2.6)

It turns out that Nature imposes some restrictions as to which configurations are to be counted as elementary events. Nature also tells us that there are two ways of listing elementary events, depending on the type of particles. For one type of particle (such as photons or ^4He atoms) called bosons, only 10 out of these configurations are to be assigned equal probability. These are shown in Fig. (2.5).

The second group of particles (such as electrons or protons), referred to as fermions, are allowed only six of these configurations. These are shown in Fig. (2.6).

In the first case (Fig. (2.5)), we say that the particles obey the Bose-Einstein statistics, and in the second case (Fig. (2.6)), we say that the particles obey the Fermi-Dirac statistics. I have brought up this example here only to show that in general, we do not have a universal rule on how to enumerate the elementary events. It is only through trial and error that one can proceed to select the elementary events, and then find whatever theoretical support there is for the eventual correct selection. In this manner, deep and profound principles of physics were discovered.[16]

2.3. Independent Events and Conditional Probability

The concepts of dependence between events and conditional probability are central to probability theory and have many uses in science.[17] In this book, we shall need only the concept of independence between two events. However, reasoning based on conditional probability appears in many applications of probability.

Two events are said to be *independent* if the occurrence of one event has no effect on the probability of occurrence of the other.

For example, if two persons who are far apart throw a fair die each, the outcomes of the two dice are independent in the sense that the occurrence of say, "5" on one die, does not have

---

[16]In Fig. (2.5), we have eliminated six configurations (within the dashed rectangle in Fig. (2.4)). All of these were counted twice in Fig. (2.4), when the particles are indistinguishable. In Fig. (2.6), we have further eliminated four more configurations (within the dashed-dotted rectangle in Fig. (2.4). For Fermion particles, two particles in one box is forbidden. (This is called the Pauli's exclusion principle.) It turns out that these rules follow from some symmetry requirements on the wave functions of the system of particles. To the best of my knowledge, the assignment of probabilities preceded the discovery of the principles of symmetry.

[17]It should be noted that the introduction of conditional probabilities and independence between events is unique to probability theory, and that is what makes probability theory differ from set theory and measure theory.

Fig. (2.7)

any effect on the probability of occurrence of a result, say, "3," on the other (left pair of dice in Fig. (2.7)). On the other hand, if the two dice are connected by an inflexible wire (right pair in Fig. (2.7)), the outcomes of the two results will be dependent. Intuitively, it is clear that whenever two events are independent, the probability of the occurrence of both events, say, "5" on one die, and outcome "3" on the other, is the *product* of the two probabilities. The reason is quite simple. By tossing two dice simultaneously, we have altogether 36 possible elementary events. Each of these outcomes has an equal probability of $1/36$, which is also equal to $1/6$ times $1/6$, i.e., the product of the probabilities of each event separately.

A second fundamental concept is conditional probability. This is defined as the probability of the occurrence of an event A given that an event B has occurred. We write this as $\Pr\{A/B\}$ (Read: Probability of A given B).[18]

Clearly, if the two events are independent, then the occurrence of B has no effect on the probability of the occurrence of A. We write that as $\Pr\{A/B\} = \Pr\{A\}$. The interesting cases occur when the events are dependent, i.e., when the occurrence of one event does affect the occurrence of the other. In everyday life, we make such estimates of conditional probabilities frequently.

Sometimes, the occurrence of one event enhances the probability of the second event; sometimes it diminishes it.

---

[18]Note that the conditional probability is defined only for a condition, the probability of which is not zero. In the abovementioned example, we require that the event B is not an impossible event.

Examples:

1) The probability that it will rain today in the afternoon, given that the sky is very cloudy at noon, is *larger* than the probability of "raining today in the afternoon."
2) The probability that it will rain today in the afternoon, given that the sky is clear at noon, is *smaller* than the probability of "raining today in the afternoon."
3) The probability that it will rain today, given that the outcome of tossing a die is "4," is the same as the probability of "raining today."

We can say that in the first example, the two events are positively correlated; in the second example, they are negatively correlated; and in the third example, they are uncorrelated or indifferent.[19]

In the three examples given above, we *feel* that the statements are correct. However, we cannot quantify them. Different persons would have made different estimates of the probabilities of "raining today in the afternoon."

To turn to things more quantitative and objective, let us consider the following events:

$A = $ {The outcome of throwing a die is "4"}

$B = $ {The outcome of throwing a die is "even"} (i.e., it is one of the following: 2, 4, 6)

$C = $ {The outcome of throwing a die is "odd"} (i.e., it is one of the following: 1, 3, 5)

---

[19]In the theory of probability, correlation is normally defined for random variables. For random variables, "independent" and "uncorrelated" events are different concepts. For single events, the two concepts are identical.

We can calculate the following two conditional probabilities:

$$\Pr\{of\ A/\ given\ B\} = 1/3 > \Pr\{of\ A\} = 1/6$$
$$\Pr\{of\ A/\ given\ C\} = 0 < \Pr\{of\ A\} = 1/6$$

In the first example, the knowledge that B has occurred *increases* the probability of the occurrence of A. Without that knowledge, the probability of A is $1/6$ (one out of six possibilities). Given the occurrence of B, the probability of A becomes *larger*, $1/3$ (one out of three possibilities). But *given* that C has occurred, the probability of A becomes zero, i.e., *smaller* than the probability of A without that knowledge.

It is important to distinguish between *disjoint* (i.e., mutually exclusive events) and *independent* events. Disjoint events are events that are mutually exclusive; the occurrence of one excludes the occurrence of the second. Being disjoint is a property of the events themselves (i.e., the two events have no common elementary event). Independence between events is not defined in terms of the elementary events comprising the two events, but in terms of their probabilities. If the two events are disjoint, then they are strongly *dependent*. The following example illustrates the relationship between dependence and the extent of overlapping.

Let us consider the following case. In a roulette, there are altogether 12 numbers $\{1, 2, 3, 4, 5, 6, 7, 8, 9, 10, 11, 12\}$

Each of us chooses a sequence of six consecutive numbers, say, I choose the sequence:

$$A = \{1, 2, 3, 4, 5, 6\}$$

and you choose the sequence:

$$B = \{7, 8, 9, 10, 11, 12\}$$

The ball is rolled around the ring. We assume that the roulette is "fair," i.e., each outcome has the same probability

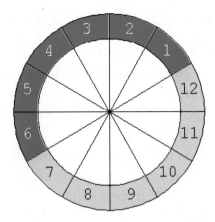

Fig. (2.8)

of $1/12$. If the ball stops in my territory, i.e., if it stops at any of the numbers I chose $\{1, 2, 3, 4, 5, 6\}$, I win. If the ball stops in your territory, i.e., if it stops at any of the numbers you chose $\{7, 8, 9, 10, 11, 12\}$, you win.

Clearly, each of us has a probability, $1/2$, of winning. The ball has an equal probability of $1/12$ of landing at any number, and each of us has 6 numbers in each territory. Hence, each of us has the same chances of winning.

Now, suppose we run this game and you are told that I won. What is the probability that you will win if you chose B? Clearly, $\Pr\{B/A\} = 0 < 1/2$, i.e., the conditional probability of B given A, is zero, which is *smaller* than the unconditional probability, $\Pr\{B\} = 1/2$. As a simple exercise, try to calculate the following conditional probabilities. In each example, my choice of the sequence $A = \{1, \ldots, 6\}$ is *fixed*. Calculate the conditional probabilities for the following *different* choices of your sequence. (Note that in this game both of us can win simultaneously.)

$$\Pr\{7, 8, 9, 10, 11, 12/A\}, \quad \Pr\{6, 7, 8, 9, 10, 11/A\},$$
$$\Pr\{5, 6, 7, 8, 9, 10/A\}, \quad \Pr\{4, 5, 6, 7, 8, 9/A\},$$

$$Pr\{3,4,5,6,7,8/A\}, \quad Pr\{2,3,4,5,6,7/A\},$$
$$Pr\{1,2,3,4,5,6/A\}.$$

Note how the correlation changes from extreme negative ("given A" *certainly* excludes your winning in the first example), to extreme positive ("given A" assures your winning in the last example). At some intermediate stage, there is a choice of a sequence that is indifferent to the information "given A." Which is the choice? If you can calculate all the abovementioned conditional probabilities, it shows that you understand the difference between disjoint events and independent events. If you cannot do the calculations, look at the answer at the end of this chapter. It is a nice exercise, although it is not essential for an understanding of the Second Law.

2.4. Three Caveats

2.4.1. *Conditional probability and subjective probability*

There is a tendency to refer to "probability" as *objective*, and to *conditional* probability as *subjective*. First, note that probability is always *conditional*. When we say that the probability of the outcome "4" of throwing a die is $1/6$, we actually mean that the *conditional* probability of the outcome "4," *given* that one of the possible outcomes: 1, 2, 3, 4, 5, 6 has occurred, or will occur, that the die is fair and that we threw it at random, and any other information that is relevant. We usually suppress this *given* information in our notation and refer to it as the unconditional probability. This is considered as an objective probability.[20]

---

[20]Here, we want to stress the *change* in the extent of "objectivity" (or subjectivity), when moving from probability of an event to a conditional probability.

Now, let us consider the following two pairs of examples:

O_1: The conditional probability of an outcome "4," given that Jacob *knows* that the outcome is "even," is $1/3$.

O_2: The conditional probability of an outcome "4," given that Abraham *knows* that the outcome is "odd," is zero.

S_1: The conditional probability that the "defendant is guilty," given that he was seen by the police at the scene of the crime, is 9/10.

S_2: The conditional probability that the "defendant is guilty," given that he was seen by at least five persons in another city at the time of the crime's commission, is nearly zero.

In all of the aforementioned examples, there is a tendency to refer to *conditional* probability as a *subjective* probability. The reason is that in all the abovementioned examples, we involved personal knowledge of the conditions. Therefore, we judge that it is highly subjective. However, that is not so. The two probabilities, denoted O_1 and O_2, are objective probabilities.

The fact that we mention the names of the persons, who are knowledgeable of the conditions, does not make the conditional probability subjective. We could make the same statement as in O_1, but with Rachel instead of Jacob. The conditional probability of an outcome "4," given that Rachel knows that the outcome is even, is $1/3$. The result is the same. The subjectivity of this statement is just an illusion resulting from the involvement of the *name* of the person who "knows" the condition. A better way of rephrasing O_1 is:

The conditional probability of an outcome "4," *given* that *we* know that the outcome is even, is $1/3$, or even better; the conditional probability of an outcome "4," *given* that the outcome is "even," is $1/3$.

In the last two statements, it is clear that the fact that Jacob or Rachel, or anyone of us *knowing* the condition does not have any effect on the conditional probability. In the last statement, we made the condition completely impersonal. Thus, we can conclude that the *given condition* does not, in itself, convert an objective (unconditional) probability into a subjective probability.

Consider the following paragraph from Callen (1983):

> *"The concept of probability has two distinct interpretations in common usage. 'Objective probability' refers to a frequency, or a fractional occurrence; the assertion that 'the probability of newborn infants being male is slightly less than one half' is a statement about census data. 'Subjective probability' is a measure of expectation based on less than optimum information. The (subjective) probability of a particular yet unborn child being male, as assessed by a physician, depends upon that physician's knowledge of the parents' family histories, upon accumulating data on maternal hormone levels, upon the increasing clarity of ultrasound images, and finally upon an educated, but still subjective, guess."*

Although it is not explicitly said, what the author implies is that in the first example: "the probability of a newborn infant being male is slightly less than one half" as stated is an answer to an unconditional probability question, "What is the probability of a newborn infant being male?" The second example is cast in the form of an answer to a conditional probability question: "What is the probability of a particular yet unborn child being male, *given*... all the information as stated."

Clearly, the answer to the second question is highly *subjective*. Different doctors who are asked this question will give *different* answers. However, the same is true for the first question if given different information. What makes the second question and its answer *subjective* is *not* the *condition* or the specific information, or the specific knowledge of this or that doctor, but the lack of sufficient knowledge. Insufficient knowledge confers liberty to give any (subjective) answer to the second question. The same is true for the first question. If all the persons who are asked have the same knowledge as stated, i.e. no information, they are free to give any answer they might have guessed. The first question is not "about the census data" as stated in the quotation; it is a question on the *probability* of the occurrence of a male gender newborn infant, *given* the information on "census data." If you do not have *any* information, you cannot answer this "objective question," but anyone given the same information on the "census data" will necessarily give the same *objective* answer.

There seems to be a general agreement that there are essentially two distinct types of probabilities. One is referred to as the judgmental probability which is highly subjective, and the second is the physical or scientific probability which is considered as an objective probability. Both of these can either be conditional or unconditional. In this book, we shall use only scientific, hence, objective probabilities. In using probability in all the sciences, we always assume that the probabilities are given either explicitly or implicitly by a given recipe on how to calculate these probabilities. While these are sometimes very easy to calculate, at other times, they are very difficult,[21] but you can

---

[21]For instance, the probability of drawing three red marbles, five blue marbles and two green marbles from an urn containing 300 marbles, 100 of each color, *given* that the marbles are identical and that you drew 10 marbles at random, and so on... This is a slightly more difficult problem, and you might not be able to calculate it, but the probability of this event is "there" in the event itself. Similarly, the probability of finding two atoms at a certain distance from each other, in a liquid at a given temperature and pressure, *given* that you know and accept the rules of statistical mechanics, and that you know that these rules have been extremely useful in predicting many average properties of macroscopic properties, etc. This probability is objective! You might not be able to calculate it, but you know it is "there" in the event.

always assume that they are "there" in the event, as much as mass is attached to any piece of matter.

2.4.2. *Conditional probability and cause and effect*

The "condition" in the conditional probability of an event may or may not be the *cause* of the event. Consider the following two examples:

1. The conditional probability that the patient will die of lung cancer, given that he or she is a heavy smoker, is 9/10.
2. The conditional probability that the patient is a heavy smoker given that he or she has lung cancer, is 9/10.

Clearly, the information given in the first condition is the *cause* (or the very probable cause) of the occurrence of lung cancer. In the second example, the information that is given in the condition — that the patient has cancer, certainly cannot be the *cause* of the patient being a heavy smoker. The patient could have started to smoke at age 20, at a much earlier time before the cancer developed.

Although the two examples given above are clear, there are cases where conditional probability is confused with causation. As we perceive causes as preceding the effect, so also is the condition perceived as occurring earlier in conditional probability.

Consider the following simple and illustrative example that was studied in great detail by Ruma Falk (1979).[22] You can view it as a simple exercise in calculating conditional probabilities. However, I believe this example has more to it. It demonstrates how we would intuitively associate conditional probability with the arrow of time, confusing causality with conditional probabilistic argument. This may or may not be relevant to the

---

[22]This example and the analysis of its implication is taken from Falk (1979).

association of the direction of change of entropy with the arrow of time (discussed further in Chapter 8).

The problem is very simple; an urn contains four balls, two white and two black. The balls are well mixed and we draw one ball, blindfolded.

First we ask: What is the probability of occurrence of the event, "White ball on first draw?" The answer is immediate: $1/2$. There are four equally probable outcomes; two of them are consistent with the "white ball" event, hence, the probability of the event is $2/4 = 1/2$.

Next we ask: What is the conditional probability of drawing a white ball on a *second* draw, given that in the first draw, we drew a white ball (the first ball is not returned to the urn). We write this conditional probability as $\Pr\{White_2/White_1\}$. The calculation is very simple. We know that a white ball was drawn on the first trial and was not returned. After the first draw, there are three balls left; two blacks and one white. The probability of drawing a white ball is simply $1/3$.

This is quite straightforward. We write

$$\Pr\{White_2/White_1\} = 1/3$$

Now, the more tricky question: What is the probability that we *drew* a white ball in the *first* draw, given that the second draw was white? Symbolically, we ask for

$$\Pr\{White_1/White_2\} = ?$$

This is a baffling question. How can an event in the "present" (white ball on the *second* draw), affect the probability of an event in the "past" (white drawn in the *first* trial)?

These questions were actually asked in a classroom. The students easily and effortlessly answered the question about $\Pr\{White_2/White_1\}$, arguing that drawing a white ball on the first draw has *caused* a change in the urn, and therefore has influenced the probability of drawing a second white ball.

However, asking about $\Pr\{White_1/White_2\}$ caused an uproar in the class. Some claimed that this question is meaningless, arguing that an event in the present cannot affect the probability of an event in the past. Some argued that since the event in the present cannot affect the probability of the event in the past, the answer to the question is $1/2$. They were wrong. The answer is $1/3$. Further discussion of this problem and Falk's analysis can be found in Falk (1979). I want to draw the attention of the reader to the fact that we are sometimes misled into associating conditional probability with cause and effect, hence we intuitively perceive the *condition* as preceding the effect; hence, the association *of conditional probability with the arrow of time.*

The distinction between *causation* and conditional probability is important. Perhaps, we should mention one characteristic property of causality that is not shared by conditional probability. Causality is transitive. This means that if *A* causes *B*, and *B* causes *C*, then *A* causes *C*. A simple example: If smoking causes cancer, and cancer causes death, then smoking causes death.

Conditional probability might or might not be transitive. We have already distinguished between positive correlation (or supportive correlation) and negative correlation (counter or anti-supportive).

If *A* supports *B*, i.e., the probability of occurrence of *B given A*, is larger than the probability of occurrence of *B*, $[\Pr\{B/A\} > \Pr\{B\}]$, and if *B* supports *C* (i.e., $[\Pr\{C/B\} > \Pr\{C\}]$), then it does not necessarily follow, in general, that *A* supports *C*.

Here is an example where supportive conditional probability is not transitive. Consider the following three events in throwing a die:

$$A = \{1, 2, 3, 4\}, \quad B = \{2, 3, 4, 5\}, \quad C = \{3, 4, 5, 6\}$$

Clearly, A supports B [i.e., [$\Pr\{B/A\} = 3/4 > P\{B\} = 2/3$]. B supports C [i.e., [$\Pr\{C/B\} = 3/4 > P\{B\} = 2/3$], but A does not support C [i.e., [$\Pr\{C/A\} = 1/2 < P\{C\} = 2/3$]

2.4.3. *Conditional probability and joint probability*

If you have never studied probability, you might benefit from reading this caveat.

I had a friend who used to ride a motorcycle. One night, while driving on the highway, he was hit by a truck and was seriously injured. When I visited him in the hospital, he was beaming and in a euphoric mood. I was sure that was because of his quick and complete recovery. To my surprise, he told me that he had just read an article in which statistics about the frequencies of car accidents were reported. It was written in the article that the chances of getting involved in a car accident is one in a thousand. The chance of being involved in two accidents in a lifetime, is about one in a million. Hence, he happily concluded: *"Now that I had this accident, I know that the chances of me being involved in another accident are very small...."* I did not want to dampen his high spirits. He was clearly confusing the probability of "having two accidents in a lifetime," with the conditional probability of "having a second accident, given that you were involved in one accident."

Of course, he might have been right in his conclusion. However, his probabilistic reasoning was wrong. If the accident was a result of his fault, then he might take steps to be very careful in the future, and might avoid driving on the highways, or at night, or stop riding a motorcycle altogether. All these will reduce his chances of getting involved in a second accident. But this argument implies that there is *dependence* between the two events, i.e., the "given condition" affects the chance of being involved in a second accident. If, however, there is no dependence between the events, say, if the accident was not his fault, even if he could be extremely careful in the future, the chances of his being involved in the next accident could not be reduced merely because he was involved in one accident!

Let us make the arguments more precise. Suppose that you tossed a coin 1000 times and all of the outcomes turned out to be heads H. What is the probability of the next outcome to be H? Most untrained people would say that chances of having 1001 heads are extremely small. That is correct. The chances are $(1/2)^{1001}$, extremely small indeed. But the question was on the *conditional* probability of a result H given 1000 heads in the last 1000 tosses. This conditional probability is one half (assuming that all the events are independent).

The psychological reason for the confusion is that you know that the probability of H and T is half. So if you normally make 1000 throws, it is most likely that you will get about 500 H and 500 T. Given that the first 1000 throws result in heads (though a very rare event) is possible, you might feel that "it is time that the chances will turn in favor of T," and the sequence of outcome must behave properly. Therefore, you feel that the chances of a tail T, given that 1000 heads have occurred, are now close to one. That is wrong, however. In fact, if a coin shows 1000 outcomes in a row to be H, I might suspect that the coin is unbalanced, and therefore I might conclude that the chances of the next H are larger than $1/2$.

To conclude, if we are given a fair coin, and it is tossed at random (which is equivalent to saying that the probability of H is $1/2$), the probability of having 1000 outcomes of H is very low $(1/2)^{1000}$, but the *conditional* probability of having the next H, given "1000-in-a-row-of-heads" is still $1/2$. This is of course true presuming that the events at each toss are independent.

2.5. A Teaspoon of Information Theory

Information theory was born in 1948.[23] "Information," like probability, is a qualitative, imprecise and very subjective concept. The same "information" obtained by different people will have different meanings, effects and values.

If I have just invested in IBM stocks while you have just sold yours, we shall have very different reactions upon reading the news that IBM has just launched a new, highly advanced computer. A farmer in Mozambique who happens to hear exactly the same news would be indifferent to this news; in fact, it might even be meaningless to him.

Just as probability theory has developed from a subjective and imprecise concept, so has information theory, which was distilled and developed into a quantitative, precise, objective and very useful theory. For the present book, information theory serves as an important milestone in the understanding of the meaning of entropy.[24]

Originally, information theory was introduced by Claude Shannon (1948) in the context of transmission of information along communication lines. Later, it was found to be very useful in statistical mechanics, as well as in many other diverse fields of research e.g. linguistics, economics, psychology and many other areas.

I shall present here only a few of the most basic ideas from information theory — the minimum knowledge necessary to use the term "information," in connection with the meaning of entropy and to answer the question, "What is the thing that

---

[23]Shannon (1948).

[24]Some authors prefer to refer to the missing information as the "uncertainty." Although I think that uncertainty is an appropriate term, my personal inclination is to prefer "information" or "missing information." I believe that in the context of the application of information theory to statistical mechanics, as cogently developed by Jaynes (1983) and Katz (1967), the word "information" is preferable.

changes?" In Chapter 8, I will argue that *entropy* is nothing but missing information as defined in information theory.

Let us start with a familiar game. I choose an object or a person, and you have to find out who or what I have chosen, by asking binary questions, i.e., questions which are only answerable by yes or no. Suppose I have chosen a person, say Einstein, you have to find out who the person is by asking binary questions. Here are the two possible "strategies" for asking questions:

Dumb "Strategy"	Smart "Strategy"
1) Is it Nixon?	1) Is the person a male?
2) Is it Gandhi?	2) Is he alive?
3) Is it me?	3) Is he in politics?
4) Is it Marilyn Monroe?	4) Is he a scientist?
5) Is it you?	5) Is he very well-known?
6) Is it Mozart?	6) Is he Einstein?
7) Is it Niels Bohr?	
8)	

I have qualified the two strategies as "dumb" and "smart." The reason is quite simple and I hope you will agree with me. The reason is as follows: If you use the first "strategy," you might of course, hit upon the right answer on the first question, while with the smart "strategy," you cannot possibly win after one question. However, hitting upon the right answer on the first guess is highly improbable. It is more likely that you will keep asking "forever," specific questions like those in the list, never finding the right answer. The reason for preferring the second "strategy" is that at each answer, you gain more *information* (see below), i.e., you exclude a large number of possibilities (ideally, half of the possibilities; see below for the more precise case). With the smart "strategy," if the answer to the first question is YES, then you have excluded a huge number of possibilities — all females. If the answer to the second question is NO, then you have excluded all living persons. In each of the additional answers you get, you narrow down further the *range* of possibilities, each time excluding a large group. With the dumb "strategy", however, assuming you are not so lucky to hit upon the right answer on the first few questions, at each point you get an answer, you exclude only *one* possibility, and practically, you almost have not changed the range of unknown possibilities. Intuitively, it is clear that by using the smart "strategy," you gain more *information* from each answer than with the dumb "strategy," even though we have not defined the term information. It seems better to be patient and choose the smart "strategy," than to rush impatiently and try to get the right answer quickly.

All that I have said above is very qualitative. That is also the reason I put the word "strategy" in quotation marks. The term "information" as used here is imprecise (to be made more precise within the framework of information theory). However, I hope you will agree with me and you will intuitively *feel* that I have correctly deemed the first list as "dumb" questions and the second as "smart." If you do not *feel* so, you should try to play the game several times, switching between the two strategies. I am sure you will find out that the smart strategy is indeed the smarter one. Shortly, we shall make the game more precise, and justify why one set of questions is deemed "dumb" and the other set "smart," and more importantly, you will have no other choice but to concur with me and be convinced, that the "smart" strategy is indeed the *smartest* one. Before doing that, however, let us ponder on this very simple game and try to understand why we cannot make any precise statement regarding the merits of the two strategies.

First, you can always argue that since you know I am a scientist, and that I am likely to choose a person like Einstein, then it would be better for you to choose the first strategy and you might even succeed. However, knowing that you know that I am a scientist, and that you might think I am likely to choose Einstein, I could outsmart you and choose Kirk Douglas instead. Perhaps, you could "out-out" smart me by knowing that I am likely to outsmart you and choose Kirk Douglas, and so on. Clearly, it is very difficult to argue along these lines.

There are many other elements of subjectivity that might enter into the game. To further illustrate this, you might have heard in this morning's news that a long sought serial killer was captured, and you might guess, or know, that I have heard the same news, and since it is still very fresh on my mind, that I am likely to choose that person.

That is why one cannot build a mathematical theory of information on this kind of game. There are too many qualitative and subjective elements that resist quantification. Nevertheless, thanks to Shannon's information theory, it is possible to "reduce" this type of game in such a way that it becomes devoid of any traces of subjectivity. Let us now describe a new game that is essentially the same as before, but in its distilled form, is much simpler and more amenable to a precise, quantitative and objective treatment.

Let us have eight equal boxes (Fig. (2.9)). I hide a coin in one of the boxes and you have to find where I hid it. All you know is that the coin *must* be in one of the boxes, and that I have no "preferred" box. Since the box was chosen at random, there is a $1/8$ chance of finding the coin in any specific box. To be neutral, the box was selected by a computer which generated a random number between 1 and 8, so you cannot use any information you might have on my personality to help you in guessing where I am "likely" to have placed the coin.

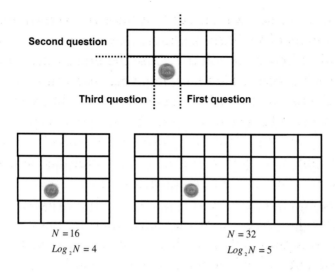

Second question

Third question : : **First question**

$N = 16$
$Log_2 N = 4$

$N = 32$
$Log_2 N = 5$

Fig. (2.9)

Note that in this game, we have completely removed any traces of subjectivity — the information we need is "where the coin is." The "hiding" of the coin can be done by a computer which chooses a box at random. You can also ask the computer binary questions to find the coin's location. The game does not depend on what you know or on what the computer knows; the required information is *there* in the game, independent of the players' personality or knowledge. We shall shortly assign a quantitative measure to this information.

Clearly, the thing you need is *information* as to "where the coin is." To acquire this information, you are allowed to ask only binary questions.[25] Instead of an indefinite number of persons in the previous game, we have only eight possibilities. More importantly, these eight possibilities have equal probabilities, $1/8$ each.

_____

[25] Of course, there are many other ways of obtaining this information. You can ask "where the coin is," or you can simply open all the boxes and see where it is. All of these are not in accordance with the rules of the game, however. We agree to acquire information only by asking *binary* questions.

Again, there are many strategies for asking questions. Here are two extreme and well-defined strategies.

The Dumbest Strategy	The Smartest Strategy
1) Is the coin in box 1?	1) Is the coin in the right half (of the eight)?
2) Is the coin in box 2?	2) Is the coin in the upper half (of the remaining four)?
3) Is the coin in box 3?	3) Is the coin in the right half (of the remaining two)?
4) Is the coin in box 4?	4) I know the answer!
5)	
⋮ ⋮	

First, note that I used the term strategy here without quotation marks. The strategies here are well-defined and precise, whereas in the previous game, I could not define them precisely. In this game, with the dumbest strategy, you ask: "Is the coin in box k," where k runs from one to eight? The smartest strategy is different: each time we divide the entire range of possibilities into two halves. You can now see why we could not define the smartest strategy in the previous game. There, it was not clear what *all* the possibilities are, and it was even less clear if division by half is possible. Even if we limited ourselves to choosing only persons who have worked in a specific field, say, in thermodynamics, we would still not know how to divide into two halves, or whether such a division is possible in principle.

Second, note that in this case, I use the adjectives "dumbest" and "smartest" strategies (I could not do that in the previous game so I just wrote "dumb" and "smart"). The reason is that here one can *prove* mathematically, that if you choose the smartest strategy and play the game many, many times, you will out beat any other possible strategies, including the worst one denoted the "dumbest." Since we cannot use the tools of mathematical proof, let me try to convince you why the "smartest" strategy is far better than the "dumbest" one (and you can also "prove" for yourself by playing this game with a friend or against a computer).

Qualitatively, if you choose the "dumbest" strategy, you might hit upon or guess the right box on the first question. But this could happen with a probability of $1/8$ and you could fail with a probability of $7/8$. Presuming you failed on the first question (which is more likely and far more likely with a larger number of boxes), you will have a chance of a right hit with a probability of $1/7$ and to a miss with a probability of $6/7$, and so on. If you miss on six questions, after the seventh question, you will *know* the answer, i.e., you will have the information as to where the coin is. If, on the other hand, you choose the "smartest" strategy, you will certainly fail on the first question. You will also fail on the second question, but you are *guaranteed* to have the required information on the third question. Try to repeat the above reasoning for a case with one coin hidden in one of 1000 boxes.

The qualitative reason is the same as in the previous game (but can now be made more precise and quantitative). By asking "Is the coin in box 1?" you might win on the first question but with very low probability. If you fail after the first question, you have eliminated only the first box and decreased slightly the number of remaining possibilities: from 8 to 7. On the other hand, with the smartest strategy, the first question eliminates *half* of the possibilities, leaving only four possibilities. The second question eliminates another half, leaving only two, and on the third question, you get the information!

In information theory, the amount of missing information, i.e., the amount of information one needs to acquire by asking questions is *defined* in terms of the distribution of probabilities.[26]

In this example, the probabilities are: $\{1/8, 1/8, 1/8, 1/8, 1/8, 1/8, 1/8, 1/8\}$. In asking the smartest question, one gains from each answer the maximum possible information (this is referred to as one bit of information). You can prove that maximum information is obtained in each question when you divide the space of all possible outcomes into two *equally probable* parts.

Thus, if at each step of the smartest strategy I gain maximum information, I will get all the information with the minimum number of questions. Again, we stress that this is true *on the average*, i.e., if we play the same game many, many times; the smartest strategy provides us with a method of obtaining the required information with the smallest number of questions. Information theory also affords us with a method of calculating the number of questions to be asked on the average, for each strategy.

Note also that the *amount* of information that is required is the same, no matter what strategy you choose. The choice of the strategy allows you to get the same amount of information with different number of questions. The smartest strategy guarantees that you will get it, on the average, by the minimum number of questions.

If that argument did not convince you, try to think of the same game with 16 boxes. I have doubled the number of boxes but the number of questions the smartest strategy needs to ask increases by only one! The average number that the dumbest strategy needs is far larger. The reason is again the same. The smartest strategy gains the maximum information at each step,

---

[26]Note that "probability" was not defined, but was introduced axiomatically. Information is *defined* in terms of probability. The general definition is: sum over all $\Pr\{i\} \log \Pr\{i\}$, where $\Pr\{i\}$ is the probability of the ith event. This has the form of an average, but it is a very special average quantity.

whereas the dumbest strategy gains little information on the first few steps. In Fig. (2.9), two more cases of the game with different numbers of boxes are shown. The number of questions in each case, calculated from information theory, is given below.[27]

The important point to be noted at this stage is that the larger the number of boxes, the greater the amount of information you would need to locate the coin, hence, the larger the number of questions needed to acquire that information. This is clear intuitively. The amount of information is *determined* by the distribution [which in our case is $\{1/N \cdots 1/N\}$ for N equally probable boxes].

To make the game completely impersonal, hence, totally objective, you can think of playing against a computer. The computer chooses a box and you ask the computer binary questions. Suppose you pay a cent for each answer you get for your binary questions. Surely, you would like to get the required information (where the coin is hidden) by paying the least amount of money. By choosing the smartest strategy, you will get maximum value for your money. In a specific definition of the units of information, one can make the amount of "information" *equal* to the number of questions that one needs to ask using the smartest strategy.

To summarize the case involving N boxes and *one* coin hidden in one of the boxes. We know that the choice of the selected box was made at random, i.e., the person who hid the coin did not have any "preference," or any bias towards any of the boxes. In other words, the box was selected with equal probability $1/N$. In this case, it is clear that the larger the number of the boxes, the larger the number of questions we need to ask to locate the hidden coin. We can say that the larger the number

---

[27]If N is the number of equally likely possibilities, then $\log_2 N$ is the number of questions you need to ask to locate the coin, e.g., for $N = 8$, $\log_2 8 = 3$; for $N = 16$, $\log_2 16 = 4$; for $N = 32$, $\log_2 32 = 5$; and so on.

of the boxes, the larger the amount of the missing information will be, and that is why we need to ask more questions.

Let us go one step further. We are told that *two* coins were hidden in N boxes. Let us assume for concreteness that the two coins were placed in two *different* boxes. Again, we know that the boxes were chosen at random. The probability of finding the first coin in a specific box is $1/N$. The probability of finding the next coin, having located the first one is only $1/(N-1)$. Clearly, in this game, we need to ask more questions to locate the two coins. In general, for a fixed number of boxes N, the larger the number of hidden coins n, the larger the number of questions we need to ask to locate them. Up to a point, when n is larger than $N/2$, we can switch to ask questions to find out which boxes are empty.[28] Once the empty boxes are found, we shall know which boxes are occupied. For $n = N$ (as for $n = 0$), we have all the information, and no questions to ask.

2.6. A Tip of a Teaspoon of Mathematics, Physics and Chemistry

As I have said in the preface, no knowledge of any advanced mathematics is needed to understand this book. If you really do not know *any* mathematics, I suggest that you just train yourself to think in terms of large numbers, very large, unimaginably large numbers. It is also useful to be familiar with the notation of exponents. This is simply a shorthand notation for large numbers. One million is written 10^6 which reads: one, followed by six zeros. Or better yet, the multiplication of ten by itself six times; $10^6 = 10 \times 10 \times 10 \times 10 \times 10 \times 10$.

---

[28]Note that we assume that the coins are identical. All we need to know is which boxes are occupied or equivalently, which boxes are empty.

For one million, it is easy to write explicitly the full number, whereas if you have a number such as 10^{23} (which is roughly the number of atoms in a one centimeter cube of a gas), you will find it inconvenient to write the number explicitly. When you get numbers of the form 10^{10000}, you will have to fill up more than a page of zeros, which is impractical! If you have a number of the form $10^{(10^{23})}$, you can spend all your life writing the zeros and still not be able to reach the end.

To get a feel for the kind of numbers we are talking about, I have just written the number "1000" in one second, i.e., one followed by three zeros. Perhaps you can write faster, say the number "10000" in one second. Suppose you are a really fast writer and you can write the number "1000000" (i.e., one followed by six zeros, which is a million) in one second; in 100 years you can write explicitly the *number* of *zeros* followed by one (assuming you will do only this)

$$6 \times 60 \times 60 \times 24 \times 365 \times 100 = 18,921,600,000$$

This is one followed by about 10^{10} zeros. This is certainly a very large number.

We can write it as:

$$10^{(10^{10})} = 10^{(1\,followed\,by\,10\,zeros)} = 10^{10000000000}$$

Suppose you or someone else did the same, not in one hundred years but for 15 billion years (about the presently estimated age of the universe), the number written explicitly would have about 10^{18} zeros, or the number itself would be

$$10^{10^{18}} = 10^{(1\,followed\,by\,18\,zeros)}$$

This is certainly an unimaginably large number. As we shall see later in this book, the Second Law deals with events that are so rare that they might "occur" once in $10^{10^{23}}$ experiments.[29] These numbers are far larger than the one you could write explicitly if you were to sit and write it for 15 billion years.

These are the kind of numbers that you will encounter when discussing the Second Law of Thermodynamics from the molecular point of view. That is all the mathematics you *need* to know to understand the Second Law. If you want to follow some of the more detailed notes, it is helpful if you could get familiar with the following three notations.

1) *Absolute value:* The absolute value of a number, written as $|x|$, simply means that whatever x is, $|x|$ is the positive value of x. In other words, if x is positive, do not change anything. If x is negative, simply delete the "minus" sign. Thus $|5| = 5$, and $|-5| = 5$. That is quite simple.

2) *Logarithm of a number.*[30] This is an extremely useful notation in mathematics. It makes writing very large numbers easy. The logarithm of a number x, is the number you must place 10^{HERE} to obtain x. Very simple! We write this as $\log_{10} x$.

Example: What is the number to be placed 10^{HERE} to obtain 1000? That is simply 3, since $10^3 = 1000$. It is written as $\log_{10} 1000 = 3$. What is the logarithm of 10000? Simply write $10000 = 10^4$ and you get $\log_{10} 10000 = 4$. This is simply the number of times you multiply 10 by itself. Although we shall not need the logarithm of any arbitrary number, it is quite clear that $\log_{10} 1975$ is larger than 3, but smaller than 4. The symbol $\log_{10}$ is called the logarithm to the base 10. In a similar fashion,

---

[29] This kind of numbers would not only take unimaginable time to write explicitly. It might be the case that such a span of time does not have a physical existence at all. Time, according to modern cosmology, could have *started* at the big-bang, about 15 billion years ago. It might *end* at the big crunch, if that event would occur in the future.

[30] We shall use only logarithm to the *base* 10. In information theory, it is more convenient to use logarithm to the base 2.

the $\log_2 x$ is the number you have put 2^{HERE} to get x. For example, the number you have to put 2^{HERE} to obtain 16 is simply 4. Since $2^4 = 16$, or equivalently, this is the number of times you multiply 2 by itself to obtain 16. One can define $\log_2 x$ for *any* positive number x, but for simplicity we shall use this notation only for x which is of the form $2^{SOME\ INTEGER}$. In information theory, the amount of information we need to locate a coin hidden in N equally likely boxes is defined as $\log_2 N$.

3) *Factorials.* Mathematicians use very useful shorthand notations for *sum* $\left(\sum\right)$ and *product* $\left(\prod\right)$. We shall not need that. There is one quite useful notation for a special product. This is denoted $N!$. It means multiplying all the numbers from 1 to N. Thus, for $N = 5$, $N! = 1 \times 2 \times 3 \times 4 \times 5$. For $N = 100$, multiply $1 \times 2 \times 3 \times \cdots \times 100$.

With these, we have covered all that we shall need in mathematics.

What about physics? As in the case of mathematics, you need not know any physics, nor chemistry, to understand the Second Law. There is, however, one fact that you should know. In Richard Feynman's lecture notes, we find the following: "*If, in some cataclysm, all of scientific knowledge were to be destroyed, and only one sentence passed on to the next generations of creatures, what statement would contain the most information in the fewest words? I believe it is the atomic hypothesis (or the atomic fact, or whatever you wish to call it) that 'all things are made of atoms — little particles that move around in perpetual motion, attracting each other when they are a little distance apart, but repelling upon being squeezed into one another.'*"

The fact is that all matter is composed of atoms (and molecules). Although nowadays this fact is taken for granted, it was not always the case. The atomic structure of matter has its roots in the philosophy of the ancient Greeks and dates back to more than two thousand years ago. It was not part of Physics,

but a philosophical speculation. There was no proof of the physical existence of atoms for almost two thousand years. Even at
the end of the nineteenth century, this hypothesis was still vig-
orously debated. When the Second Law of Thermodynamics
was formulated, the atomic nature of matter was far from being
an established fact. Boltzmann was one of the promoters of the
atomic structure of matter, and as we have noted, he opened the
door to a molecular understanding of the Second Law of Thermodynamics. He had to face powerful opponents who claimed
that the existence of the atom was only a hypothesis and that
the atomic structure of matter was a speculation and therefore
should not be a part of physics. Nowadays, the atomic structure
of matter is an accepted fact.

The second fact that you should know is somewhat more
subtle. It is the indistinguishability of atoms and molecules. We
shall spend a considerable time playing with dice, or coins. Two
coins could be different in color, size, shape, etc., such as shown
in Fig. (2.10).

In daily life, we use the terms identical and indistinguishable
as synonyms. The two coins in Fig. (2.11) are identical in shape,
size, color, and in anything you like.

Fig. (2.10)

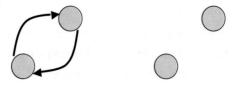

Fig. (2.11)

We shall say that they are *distinguishable* from each other, in the sense that if they move around, we can follow each of the coins with our eyes, and at any given time, we can tell which coin came from which place.

Suppose you interchange the two coins on the left of (Fig. (2.11)) to obtain the configuration on the right. If the coins are identical, then you cannot tell the difference between the left and the right configurations. However, if you follow the process of interchanging the coins, you can tell where each coin came from. This is impossible to tell, in principle, for identical molecular particles. For these cases, we use the term indistinguishability rather than identity.

Consider a system of two compartments (Fig. (2.12)). Initially, we have 5 coins in each compartment which are separated by a partition. The coins are identical in all aspects. We now remove the partition and shake the whole system. After some time we shall see a new configuration like the one on the right side.

We say that the particles (in this case, the coins) are distinguishable even though they are identical, if we can tell at any

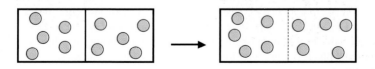

Fig. (2.12)

point of time which coin originated from which compartment, simply because we can follow the trajectories of all the particles. We say that the particles are indistinguishable when we do the same experiment as above, but we cannot tell which is which after removing the partition. We cannot follow the trajectories of the particles. This fact was foreign to classical mechanics. In classical mechanics, we always think of particles of any size as being "labeled," or at *least*, are in *principle*, "labelable." Newton's equation of motion predicts the trajectories of each specific particle in the system. In quantum mechanics, this is not possible, in *principle*.[31]

In the next two chapters, we shall have plenty of examples where we start with distinguishable dice, and *voluntarily* disregard their distinguishability. This process of "un-labeling" the particles will be essential in understanding the Second Law of Thermodynamics.

As for chemistry, you do not need to know anything beyond the fact that matter is composed of atoms and molecules. However, if you want to understand the example I have given in the last part of Chapter 7 (interesting, but not essential), you need to know that some molecules have an asymmetrical center (or chiral center) and these molecules come in pairs, denoted by l and d;[32] they are almost identical, but one is a mirror image of the other (these are called enantiomers). All amino acids — the building blocks of proteins, that are in turn the building

---

[31]Note that we can distinguish between *different* and *identical* particles. We cannot distinguish by any of our senses between *identical* and *indistinguishable* particles (that is why they are considered colloquially as synonyms). Note also that we can change from *different* to *identical* particles continuously, at least in theory. But we cannot change from *identical* to *indistinguishable*, in a continuous way. Particles are either distinguishable or indistinguishable. Indistinguishability is not something we observe in our daily life. It is a property imposed by Nature on the particles.

[32]The two isomers rotate a polarized light in different directions l (for levo) to the left and d (for dextro) to the right.

$$
\begin{array}{cc}
\overset{\displaystyle H}{\underset{\displaystyle NH_2}{\text{HOOC}-\text{C}-\text{CH}_3}}
&
\overset{\displaystyle H}{\underset{\displaystyle NH_2}{\text{CH}_3-\text{C}-\text{COOH}}}
\\[2em]
\textit{Alanine l} & \textit{Alanine d}
\end{array}
$$

Fig. (2.13)

blocks of muscles and many other tissues of the body — come in one version of these pairs. An example is the amino acid alanine (Fig. (2.13)).

Most of the natural amino acids are of the *l-form*; they are distinguished by their optical activity. The only thing you need to know is that these pairs of molecules are almost exactly the same as far as the molecular parameters such as mass, moment of inertia, dipole movement, etc. are concerned. Molecules of the same type, say the *l-form*, are indistinguishable among themselves, but the *l-form* and the *d-form* are distinguishable, and in principle, are separable.

At this stage, we have discussed all the required "pre-requisites" you need for understanding the rest of the book. Actually, the only pre-requisite you need in order to follow the pre-requisites given in this chapter is common sense. To test yourself, try the following two quizzes.

2.7. A Matter of Lottery

A state lottery issued one million tickets. Each ticket was sold at $10; therefore, the gross sales of the dealer were $10,000,000. One number wins $1,000,000. There were 999,000 numbers, each winning a prize, the value of which is one US$ or its equivalent, but in *different* currencies. There are also 999 prizes, each with a $10 value, but in *different* currencies. So altogether, the dealers must distribute the equivalent of US$2,008,990. This leaves a nice, net gain of about $8,000,000 to the dealers. Note that all the one dollar gains and the ten dollar gains are in *different* currencies and *distinguishable*. In other words, two persons gaining a one dollar *value*, get *different* dollars, i.e., different prizes having the same value.

Now the questions:

Answer the first three questions by Yes or No.
 I bought only one ticket, and

1) I tell you that I won $1,000,000 in the lottery, would you believe it?
2) I tell you that I won a one-dollar equivalent in Indian currency, would you believe it?
3) I tell you that I won a ten-dollar equivalent in Chinese currency, would you believe it?

   Now, estimate the following probabilities:

4) What is the probability of winning the $1,000,000 prize?
5) What is the probability of winning the one dollar in Indian currency?
6) What is the probability of winning the ten dollar equivalent in Chinese currency?
7) What is the probability of winning a one dollar *valued* prize?
8) What is the probability of winning a ten dollar *valued* prize?
9) After answering questions 4–8, would you revise your answers for questions 1–3?

For answers, see the end of the chapter (Section 2.10.2).

2.8. A Matter of Order-Disorder

Look at the two figures on the first page of the book. I named them ordered and disordered. Let us refer to them as B and A.[33] Now that you know some elementary notions of probability, try to answer the following questions. Suppose that 200 dice were used to construct each of the figures A and B.

1. I tell you that I got the exact two configurations A and B by throwing 200 dice twice on the board. Would you believe it?
2. I tell you that I arranged configuration A, then I shook the table for a few seconds and got B. Would you believe it?
3. I tell you that I arranged configuration B, then I shook the table for a few seconds and got A. Would you believe it?

    To answer the abovementioned questions, you need only to make a qualitative estimate of the likelihood of the two events A and B. Now, let us examine if you can also make quantitative estimates of probabilities. Suppose I tell you that the board is divided into, say, 1000 small squares. Each square can contain only one die. The orientation of the die is unimportant, but the upper face of each die is important. A configuration is an exact specification of the upper face and location (i.e., in which square) of each die. There are altogether 200 dice (do not count the dice, nor the squares in the figure; they are different from 200 and 1000, respectively). Each die can have one of the six numbers $(1, 2 \ldots 6)$ on its upper face, and it can be located in any one of the 1000 little squares (but not more than one die in a single square, and orientation does not matter). Now, estimate the probabilities of the following events:

4. The probability of obtaining the *exact* configuration A
5. The probability of obtaining the *exact* configuration B

---

[33]Tentatively, B for Boltzmann and A for Arieh.

6. The probability of obtaining the exact configuration *A* but now disregarding the number of dots on the dice.
7. The probability of obtaining the exact configuration *B* but now disregarding the number of dots on the dice.

After practicing with these (tiny) probabilities and presuming you have answered correctly, try the next two questions. They are the easiest to answer.

8. I threw the same 200 dice twice a second time, and I got exactly the same two configurations *A* and *B*. Would you believe it?

Now look carefully at configuration *A*. Do you see any recognizable patterns or letters? Look only at the dice that have only one dot on their faces. Do you recognize any patterns? If you do not, look at the same figure on the last page of the book (you should now understand why I chose A, for Arieh and B, for Boltzmann). Now, the last question:

9. Would you revise any of your answers for the 8 questions above?

Answers are given at the end of this chapter (Section 2.10.3).

2.9. A Challenging Problem

The following is a problem of significant historical value. It is considered to be one of the problems, the solution of which has not only crystallized the concept of probability, but also transformed the reasoning about chances made in gambling salons into mathematical reasoning occupying the minds of mathematicians.

Neither the problem nor its solution is relevant to an understanding of the Second Law. My aim in telling you this story is three-fold. First, to give you a flavor of the type of problems which were encountered at the time of the birth of the probability theory. Second, to give you a flavor of the type of difficulties encountered in calculating probabilities, even in problems that seem very simple. And finally, if you like "teasing" problems, you will savor the delicious taste of how mathematics can offer an astonishingly simple solution to an apparently difficult and intractable problem.

The following problem was addressed to Blaise Pascal by his friend Chevalier de Mere in 1654.[34]

Suppose two players place $10 each on the table. Each one chooses a number between one to six. Suppose Dan chose 4 and Linda chose 6. The rules of the game are very simple. They roll a single die and record the sequence of the outcomes. Every time an outcome "4" appears, Dan gets a point. When a "6" appears, Linda gets a point. The player who collects three points first wins the total sum of $20. For instance, a possible sequence could be:

$$1, 4, 5, 6, 3, 2, 4, 6, 3, 4.$$

Once the number 4 appears three times, Dan wins the entire sum of $20.

Now, suppose the game is started and at some point in time the sequence of the outcomes is:

$$1, 3, 4, 5, 2, 6, 2, 5, 1, 1, 5, 6, 2, 1, 5$$

At this point, there is some emergency and the game must be stopped! The question is how to divide the sum of $20 between the two players.

Note that the problem does not arise if the rules of the game explicitly instruct the player on how to divide the sum should

---

[34]The historical account of these on other earlier probabilistic problems can be found in David (1962).

the game be halted. But in the absence of such a rule, it is not clear as to how to divide the sum.

Clearly, one feels that since Dan has "collected" one point, and Linda has "collected" two, Linda should get the bigger portion of the $20. But how much bigger? The question is, what is the *fairest* way of splitting the sum, given that sequence of outcomes? But what do we mean by the *fairest*, in terms of splitting the sum? Should Linda get twice as much as Dan because she has gained twice as many points? Or perhaps, simply split the sum into two equal parts since the winner is undetermined? Or perhaps, let Linda collect the total sum because she is "closer" to winning than Dan.

A correspondence between Blaise Pascal and Pierre de Fermat ensued for several years. These were the seminal thoughts which led to the development of the theory of probability. Note that in the 17th century, the concept of probability still had a long way to go before it could be crystallized. The difficulty was not only of finding the mathematical solution. It was no less difficult to clarify what the problem was, i.e., what does it mean to find a *fair* method of splitting the sum?

The answer to the last question is as follows:

As there was no specific rule on how to divide the sum in case of halting of the game, the "fairest" way of splitting the sum would be to divide it according to the ratio of the probabilities of the two players in winning the game had the game been continued.

In stating the problem in terms of probabilities, one hurdle was overcome. We now have a well-formulated problem. But how do we calculate the *probabilities* of either player winning? We feel that Linda has a better chance of winning since she is "closer" to collecting three points than Dan. We can easily calculate that the probability that Dan will win, on the *next* throw is *zero*. The probability of Linda winning on the next throw is $1/6$, and the probability of neither one of them winning on the next throw is $5/6$. One can calculate the probabilities of winning after two throws, three throws, etc., It gets very complicated, and in principle, one needs to sum over an infinite series, so the mathematical solution to the problems in this way is not easy. Try to calculate the probability of each player winning on the next two throws, and on the next three throws, and see just how messy it can get. Yet, if you like mathematics, you will enjoy the simple solution based on solving a one-unknown-equation given at the end of this chapter (Section 2.10.4).

2.10. Answers to the Problems

2.10.1. *Answers to the roulette problems*

In all of these problems, my choice of a sequence is fixed: $\{1, 2, 3, 4, 5, 6\}$; it has the probability of $1/2$ of winning. If you choose the *disjoint* event $\{7, 8, 9, 10, 11, 12\}$, then the conditional probability is

$$\Pr\{B/A\} = \Pr\{7, 8, 9, 10, 11, 12/A\} = 0$$

This is because knowing that A occurs excludes the occurrence of B. In the first example of choosing an overlapping sequence, we have (see Fig. (2.14))

$$\Pr\{B/A\} = \Pr\{6, 7, 8, 9, 10, 11/A\} = 1/6 < 1/2$$

Knowing that A has occurred means that your winning is possible only if the ball landed on "6"; hence, the conditional probability is $1/6$, which is smaller than $\Pr\{B\} = 1/2$, i.e., there is negative correlation.

Similarly for $B = \{5, 6, 7, 8, 9, 10\}$, we have

$$\Pr\{B/A\} = \Pr\{5, 6, 7, 8, 9, 10/A\} = 2/6 < 1/2$$

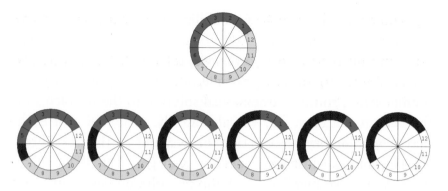

Fig. (2.14)

Here, "given A," you will win only if the ball lands on either "5" or "6"; hence, the conditional probability is $2/6$, which is still smaller than $\Pr\{B\} = 1/2$.

In the third case, $B = \{4, 5, 6, 7, 8, 9\}$, hence

$$\Pr\{B/A\} = \Pr\{4, 5, 6, 7, 8, 9/A\} = 3/6 = 1/2$$

Here, the conditional probability is $1/2$, exactly the same as the "unconditional" probability $\Pr(B) = 1/2$, which means that the two events are *independent*, or uncorrelated.

For the last three examples, we have

$$\Pr\{B/A\} = \Pr\{3, 4, 5, 6, 7, 8/A\} = 4/6 > 1/2$$
$$\Pr\{B/A\} = \Pr\{2, 3, 4, 5, 6, 7/A\} = 5/6 > 1/2$$
$$\Pr\{B/A\} = \Pr\{1, 2, 3, 4, 5, 6/A\} = 6/6 = 1 > 1/2$$

In the last example, knowing that A occurs makes the occurrence of B certain. In these examples, we have seen that overlapping events can be either positively correlated, negatively correlated, or non-correlated.

2.10.2. *Answer to "a matter of lottery"*

1) You will probably not believe it, although it is possible.

2) You will probably believe it, although the chances are as low as in winning the $1,000,000 prize.
3) You will probably believe it, although the chances are as low as in winning the $1,000,000 prize.
4) The probability is one in a million (10^{-6}).
5) The probability is one in a million (10^{-6}).
6) The probability is one in a million (10^{-6}).
7) The probability is $999000/1000000 \approx 1$.
8) The probability is $999/1000000 \approx 1/1000$.
9) If your answer to 1 was NO, you are quite right. The chances are very low indeed. If your answers to 2 and 3 were YES, you are probably wrong. The chances are as low as winning the $1,000,000 prize.

If your answers to questions 2 and 3 were YES, you are probably confusing the *exact* event "winning a dollar in a *specific* currency" with the *dim* event,[35] "winning a dollar in *any* currency." The first is a very unlikely event while the second is almost certain.

2.10.3. *Answer to "a matter of order-disorder"*

1) You should probably believe A but not B. But see below.
2) You should not believe that.
3) You might believe that, if you consider A as one of a randomly obtained configuration. But you should not believe it if you consider the *specific* configuration A.
4) The probability of one die showing a specific face *and* being in a specific location is $\frac{1}{6} \times \frac{1}{1000}$. The probability of all the 200 dice showing specific faces and being in specific locations (note that the dice are distinguishable and that no more than

---

[35]The term "*dim* event" will be discussed in the next chapters.

one die in one box is allowed) is $\left(\frac{1}{6}\right)^{200} \frac{1}{1000\times999\times998\times\cdots\times801}$. This is a very small probability.

5) The same as in question 4.

6) The probability is $1/1000 \times 999 \times 998 \times \cdots \times 801$, still very small.

7) The probability is as in question 6, still very small.

8) You probably should not. You might be tempted to believe that I got configuration A presuming it is a random configuration. However, the question refers to the *exact* configuration A.

9) Make sure that whenever the question applies to a *specific* configuration like A or B, the probability is extremely small. However, if you consider configuration A as a random configuration, you might be right in assigning to this configuration a larger probability. The reason is that there are many configurations that "look" the same as A, i.e., a random configuration, hence, a large probability. However, after realizing that A contains the word "Arieh," it is certainly not a random configuration.

2.10.4. *Answer to "a challenging problem"*

The solution to the problem is this. Denote by X the probability of Linda winning. Now, in the *next* throw, there are three mutually exclusive possibilities:

I: outcome $\{6\}$ with probability $^1/_6$
II: outcome $\{4\}$ with probability $^1/_6$
III: outcome $\{1, 2, 3, 5\}$ with probability $^4/_6$

Let us denote the event "Linda wins" by LW. The following equation holds

$$X = \Pr(LW) = \Pr(I)\Pr(LW/I) + P(II)\Pr(LW/II)$$
$$+ \Pr(III)\Pr(LW/III) = {}^1/_6 \times 1 + {}^1/_6 \times {}^1/_2 + {}^4/_6 \times X$$

This is an equation with one unknown $6X = 3/2 + 4X$. The solution is $X = 3/4$.

Note that the events I, II, and III refer to the possible outcomes on the *next* throw. The event "LW" refers to "*Linda wins*," regardless of the number of subsequent throws. The equation above means that the probability of Linda winning is the sum of the three probabilities of the three mutually exclusive events. If event I occurs, then she wins with probability one. If event II occurs, then she has probability $1/2$ of winning. If event III occurs, then the probability of her winning is X, the same probability as at the time the game was halted.

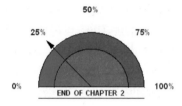

END OF CHAPTER 2

First Let Us Play with Real Dice

3.1. One Die

We will start with a very dull game. You choose a number between 1 and 6, say "4," and I choose a different number between 1 and 6, say "3." We throw the die. The first time the result 4 or 3 appears, either you or I win, respectively. This game does not require any intellectual effort. There is no preferred outcome; each outcome has the same likelihood of appearing and each one of us has the same chance of winning or losing. If we play the same game many times, it is likely that on the average, we shall end up even, not gaining nor losing (presuming the die is fair, of course). How do I know that? Because we have accepted the fact that the probability of each outcome is $1/6$ and we also believe, based on our experience, that no one can beat the laws of probability. However, this was not always so. In earlier times, it was believed that some people had divine power which enabled them to predict the outcome of throwing a die, or that there existed some divine power that determined the outcomes at it His will. So if we could communicate with "Him," either directly or through a mediator, we might know better what outcome to choose.[1] Today, however, when we consider the same game of die, we assume that there are six, and only six possible outcomes (Fig. (3.1)) and each of the outcome

---

[1] See footnote on page 20, Chapter 2.

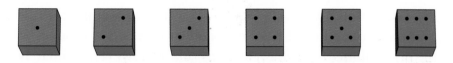

Fig. (3.1)

has the same probability, $^1/_6$. The next table is trivial.

Outcome:	1	2	3	4	5	6
Probability:	$^1/_6$	$^1/_6$	$^1/_6$	$^1/_6$	$^1/_6$	$^1/_6$

3.2. Two Dice

A slightly more complicated game is to play with two dice. There are different ways of playing with two dice. We could choose, for example, a *specific* outcome say, "white die, 6 and blue die, 1." There are altogether 36 possible *specific* outcomes; these are listed below.

1.1,	1.2,	1.3,	1.4,	1.5,	1.6
2.1,	2.2,	2.3,	2.4,	2.5,	2.6
3.1,	3.2,	3.3,	3.4,	3.5,	3.6
4.1,	4.2,	4.3,	4.4,	4.5,	4.5
5.1,	5.2,	5.3,	5.4,	5.5,	5.6
6.1,	6.2,	6.3,	6.4,	6.5,	6.6

Clearly, these are all equally likely outcomes. How do I know? Assuming that the dice are fair and the outcomes are independent (one outcome does not affect the other), then the answer follows from plain common sense. Alternative answer: each outcome of a single die has probability $^1/_6$. Each specific outcome of a pair of dice is the product of the probabilities of each die, i.e., $1/6$ times $1/6$ which is $1/36$. This argument requires the rule that the probability of two independent events is a product of the probabilities of each event. This rule is, however, ultimately based on common sense as well!

As in the case of one die, playing this game is dull and uninteresting, and certainly does not require any mental effort.

A slightly more demanding game is to choose the *sum* of the outcomes of the two dice, regardless of the specific numbers or the specific colors of the dice. Here are all the possible outcomes in this game.

Outcome: 2,   3,   4,   5,   6,   7,   8,   9,   10,   11,   12

Altogether, we have eleven possible outcomes. We shall refer to these as dim events for reasons explained below.[2] If you have to choose an outcome, which one will you choose? Contrary to our two previous games, here you have to do a little thinking. Not much and certainly within your capability.

Clearly, the outcomes listed above are *not* elementary events, i.e., they are not equally probable. As you can see from Fig. (3.2a), or count for yourself, each event consists of a sum (or union) of elementary events. The elementary events of this game are the same as in the previous game, i.e., each specific outcome has the probability of $1/36$. Before calculating the probabilities of the compound events, i.e., the events having a specific sum, take note that in Fig. (3.2a), the events of equal sum feature along the principal diagonal of the square. This figure is reproduced in Fig. (3.2b) rotated by 45° (clockwise). Once you realize that you can count the number of *specific* (or elementary) events contained in each of the compound events, you can now calculate the probability of each of these events. To facilitate

---

[2]Here, we use the term dim-event to refer to an event; the details of the events comprising it are disregarded.

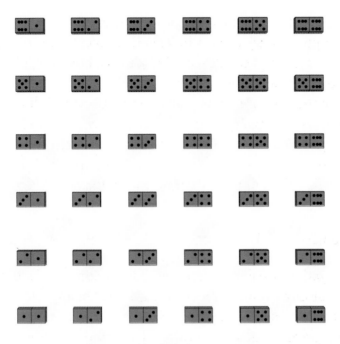

Fig. (3.2a)

counting, we have "compressed" the rotated Fig. (3.2b) to produce Fig. (3.2c) (each pair is rotated back counter-clockwise), and regrouped the pairs with equal sum.

Dim Events	2	3	4	5	6	7	8	9	10	11	12
Multiplicity	1	2	3	4	5	6	5	4	3	2	1
Probability	$1/36$	$2/36$	$3/36$	$4/36$	$5/36$	$6/36$	$5/36$	$4/36$	$3/36$	$2/36$	$1/36$

The probabilities of the compound events are listed in the table above. The "multiplicity" is simply the number of *specific* events comprising the *dim* event.

How do I know that these are the *right* probabilities? The answer is again pure common sense. You should convince yourself of that. If you are not convinced, then play this game a

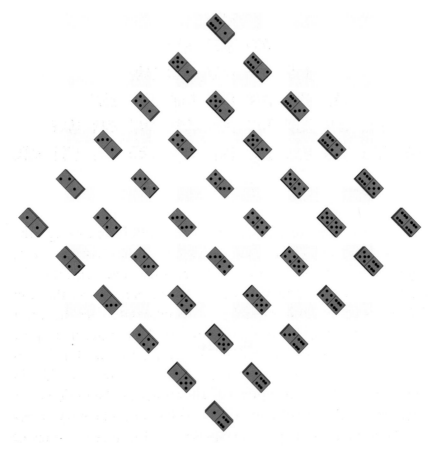

Fig. (3.2b)

few million times and record the frequencies of the outcomes. However, I will advise you not to do the experiment, but instead, *trust* your common sense to lead you to these probabilities, or equivalently make a million *mental experiments* and figure out how many times each sum will occur. Once you are convinced, check that the probabilities of all the events sum up to one, as it should be.

With this background on the game of two dice, let us proceed to play the game. Which outcome will you choose? Clearly, you will not choose 2 or 12. Why? Because, these events consist of

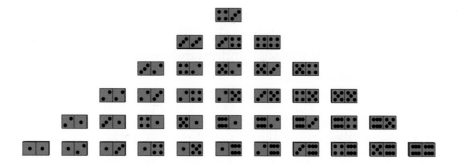

Fig. (3.2c)

only one specific (or elementary) event. Looking through the rotated table (Fig. (3.2b)), you will see that your best chance of winning is to choose outcome 7. There is nothing magical in the number 7. It so happens that in this particular game, the sum 7 consists of the *largest* number of specific outcomes; hence, it is the number that is most likely to win. Of course, if we play the game only once, you might choose 2 and win. But if you choose 2 and I choose 7, and if we play many times, I will win most of the time. The relative odds are 6:1 as the table above indicates. In Fig. (3.3), we plot the number of elementary events (or the number of specific configurations) for different sums of the games with one and two dice.

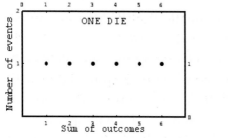
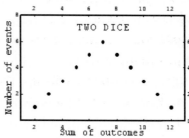

Fig. (3.3)

Once you feel comfortable with this game, we can proceed to the next game. A slightly more difficult one, but it puts you on the right track towards understanding the Second Law of Thermodynamics.

3.3. Three Dice

This game is essentially the same as the previous one. It is a little more difficult, and entails more counting. We play with three dice and we have to choose the *sum* of the outcome, only the sum, regardless of the specific numbers on each die or its color. There is nothing new, in principle, only the counting is more tedious. Indeed, this is exactly the type of game from which the theory of probability evolved. Which outcome to choose? That question was addressed to mathematicians before the theory of probability was established (see Chapter 2).

The list of all possible outcomes is:

3, 4, 5, 6, 7, 8, 9, 10, 11, 12, 13, 14, 15, 16, 17, 18

Altogether, there are 16 different outcomes. To list all the possible specific outcomes, e.g., $\{blue = 1, red = 4, white = 3\}$, would take quite a big space. There are altogether $6^3 = 216$ possible specific outcomes. You clearly would not like to bet on 3 or 18, neither on 4 nor on 17. Why? For exactly the same reason you have not chosen the smallest or the largest numbers in the previous game. But what *is* the best outcome to choose? To answer this question, you have to count all the possible specific (or elementary) outcomes that give rise to each of the sums in the list above. This requires a little effort, but there is no new principle involved, only common sense and a willingness to do the counting. Today, we are fortunate to have the computer to do the counting for us. The results are listed in the table below.

In the second row, we list the number of possibilities for each sum. The probabilities are obtained from the second row by dividing through by 216.

Sum	3	4	5	6	7	8	9	10	11	12	13	14	15	16	17	18
Multiplicity	1	3	6	10	15	21	25	27	27	25	21	15	10	6	3	1
Probability	$\frac{1}{216}$	$\frac{3}{216}$	$\frac{6}{216}$	$\frac{10}{216}$	$\frac{15}{216}$	$\frac{21}{216}$	$\frac{25}{216}$	$\frac{27}{216}$	$\frac{27}{216}$	$\frac{25}{216}$	$\frac{21}{216}$	$\frac{15}{216}$	$\frac{10}{216}$	$\frac{6}{216}$	$\frac{3}{216}$	$\frac{1}{216}$

Thus, for one die, the distribution is uniform. For two, we find a maximum at *sum* = 7 (Fig. (3.3)). For three dice, we have two maximal probabilities at the sums of 10 and 11. The probabilities of these are $^{27}/_{216}$. Therefore, if you want to win, choose either sum 10 or sum 11. In Fig. (3.4), we plot the number of elementary events for each possible sum of the three dice. By dividing the number of events by the total number of specific events $6^3 = 216$, we get the corresponding probabilities which are also plotted in Fig. (3.4).

You should be able to confirm for yourself some of these numbers. You do not need any higher mathematics, nor any knowledge of the theory of probability — simple counting and common sense are all you need. If you do that, and if you understand why the maximum probabilities occur at 10 or 11, you are almost halfway through to understanding the Second Law.

Let us proceed to discuss a few more games of the same kind but with an increasing number of dice.

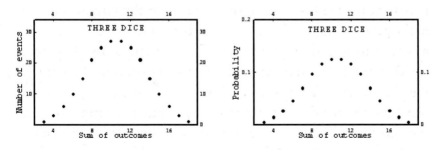

Fig. (3.4)

3.4. Four Dice and More

The four-dice game is the same as before. We have to choose a number from 4 to 24. We throw four dice simultaneously and look at the sum of the four outcomes on the faces of the four dice, regardless of the identity (or the color), or specific number on the face of the specific die; only the sum matters.

In this case, we have $6^4 = 1296$ possible specific outcomes. It is not practical to list all of these. The probability of each *specific* outcome is $1/1296$. The counting in this case is quite laborious but there is no new principle. Figure (3.5) shows the probabilities, i.e., the number of specific outcomes, divided by the total number of specific outcomes, as a function of the sum. For four dice, the range of possible sums is from the minimum 4 to the maximum 24. For five dice, the range is from 5 to 30. For six dice, the range is from 6 to 36 and for seven dice, the range is from 7 to 42.

In Fig. (3.6), we plot the same data as in Fig. (3.5), but this time we plot the probabilities as a function of the "reduced" sum. The reduced sum is simply the sum divided by the maximum sum. Thus, the different ranges from 0 to N in Fig. (3.5),

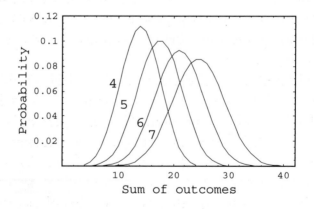

Fig. (3.5)

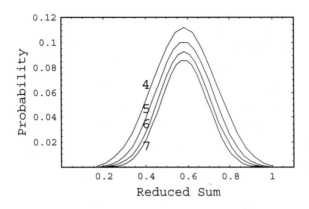

Fig. (3.6)

are all "compressed" into the same range — from 0 to 1. This compression changes the area under the curves. In Fig. (3.5), the area under each curve is one, whereas in Fig. (3.6), it is reduced by the factor N. Note that the spread of the probabilities in Fig. (3.5) is larger as N increases. In contrast, the spread in Fig. (3.6) is diminished with N. The larger N, the sharper the curve. This means that if we are interested in *absolute* deviations from the maximum (at $N/2$), we should look at Fig. (3.5). However, if we are only interested in the *relative* deviations from the maximum (at $1/2$), we should look at the curves in Fig. (3.6). When N becomes very large, the curve in Fig. (3.6) becomes extremely sharp, i.e., the relative deviations from the maximum becomes negligibly small.

Take note also that in each case, there are either one or two sums for which the probability is maximum. Observe how the shape of the distribution turns out. This resembles the bell shape, also known as the Normal distribution or the Gaussian distribution. This is an important shape of a probability distribution in the theory of probability and statistics, but it is of no concern to us here. You will also note that as the number of dice increases, the curve in Fig. (3.6) becomes narrower, and the maximum

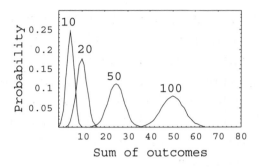

Fig. (3.7)

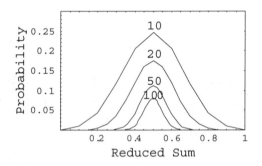

Fig. (3.8)

probability becomes lower. The lowering of the maxima is the same as in Fig. (3.5), but the "spread" of the curve is different. We shall further discuss this important aspect of the probabilities in the next chapter. In Figs. (3.7) and (3.8), we show similar curves as in Figs. (3.5) and (3.6) but for larger values of N.

We stop at this stage to ponder on what we have seen so far, before proceeding to the next step. You should ask yourself two questions. First, which the winning number in each case is; and second, the more important question, why this is a winning number. Do not worry about the exact counting; just be assured that in each game there are one or two sums with the largest probability. This means that if you play this game many times, these specific sums will occur more frequently than all the other sums, and if you are really going to play this game, you would better choose one of these winning numbers.

The first question is important, even critical, if you are interested in *playing* this game. However, if you want to understand the Second Law and to be able to follow the arguments in the chapters that follow, you should ponder on the question "Why." Why is there such a winning number? Let us take the case of three dice. Think of the reasons for the *existence* of such a winning number. Start with the *sum* = 3. We have only *one* specific outcome; here it is:

$$blue = 1, \quad red = 1, \quad white = 1 \ldots \quad sum = 3$$

You can imagine that this *specific* outcome will be a very rare event. So is the event, *sum* = 18. There is only one *specific* outcome that gives the sum 18. Here it is:

$$blue = 6, \quad red = 6, \quad white = 6 \ldots \quad sum = 18$$

For the sum = 4, we have *three* specific outcomes. Here they are:

$$blue = 1, \quad red = 1, \quad white = 2 \ldots \quad sum = 4$$
$$blue = 1, \quad red = 2, \quad white = 1 \ldots \quad sum = 4$$
$$blue = 2, \quad red = 1, \quad white = 1 \ldots \quad sum = 4$$

The partition of the *sum* = 4 into three integers is unique 1:1:2, and if we do not care which die carries the outcome 1, and which carries 2, we shall not distinguish between the three cases listed above. We shall refer to *each* of the three abovementioned possibilities as a *specific* configuration. If we disregard the differences in the specific configuration and are only interested in the *sum* = 4, we shall use the term *dim* configuration or *dim* event.

Next, let us look at *sum* = 5. Here we have six *specific* possibilities.

Blue = 1,   Red = 1,   White = 3... *sum* = 5
Blue = 1,   Red = 3,   White = 1... *sum* = 5
Blue = 3,   Red = 1,   White = 1... *sum* = 5
Blue = 1,   Red = 2,   White = 2... *sum* = 5
Blue = 2,   Red = 2,   White = 1... *sum* = 5
Blue = 2,   Red = 1,   White = 2... *sum* = 5

Here, we have two causes for the multiplicity of the outcomes; first, we have different partitions (1:1:3 and 2:2:1), and each partition comes in three different combinations of colors, i.e., each partition has weight 3. We shall say that there are six *specific* configurations, but only one *dim* configuration or *dim* event.

The terms dim event and specific event are important to an understanding of the Second Law. As we have seen, each of the *specific* events has the same probability. In the two-dice game, a specific event is a specific list of *which* die contributes which number to the total sum. For the case of three dice, if we disregard the color of the dice and the specific contribution of each die to the total sum, then we find that for *sum* = 3, we have one dim event consisting of only one specific event. For *sum* = 4, we have one dim event consisting of three specific events, for *sum* = 5, we have one dim event consisting of six specific events, and so on.

In the next chapter, we shall play a modified game of dice which will bring us closer to the real experiment to be discussed in Chapter 7. We shall play with more primitive dice, three surfaces of each bearing the number "0" and the other three bearing the number "1." The simplification in switching from real dice to the simplified dice is two-fold. First, we have only two out-comes for each die (zero or one); and second, for each sum we have only *one* partition. The number of specific outcomes com-prising the dim event *sum* = *n* is simply *n*, i.e., the number of faces showing "1."

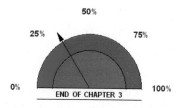

END OF CHAPTER 3

Let's Play with Simplified Dice and have a Preliminary Grasp of the Second Law

The new game is simpler than the previous one. We consider either dice with three faces marked with "0," and three faces marked with "1," or coins with one side marked with "0" and the other side marked "1." Since we started with dice, we shall continue with these, but you can think in terms of coins if you prefer. The important thing is that we do an "experiment" (tossing a die or a coin), the outcomes of which are either "0" or "1," with equal probabilities $1/2$ and $1/2$. This is one simplification. Instead of six possible outcomes, we have now only two. The second simplification arises from the specific choice of "0" and "1." When we *sum* the outcomes of N dice, the sum is simply the *number* of ones. The zeros do not add to the counting. For instance, with $N = 10$ (Fig. (4.1)), we might have a specific outcome 1,0,1,1,0,1,0,1,0,0. The *sum* is 5. It is also the *number* of "ones" in this outcome (or the total number of dots in the dice having a zero or one dot on their faces).

We shall be interested in the "evolution" of the game and not so much with the winning strategies. But if you are more comfortable with a gambling game, here are the rules.

Fig. (4.1)

We play with N fair dice, with "0" and "1" outcomes only. We always start with a pre-arranged *configuration*; all the dice show zero.[1]

By configuration, we mean the detailed specification of the sequence of "zeros" and "ones," e.g., first die shows "1," second die shows "0," third shows "0," etc. We shall refer to the initial configuration as the *zeroth* step of the game.

For $N = 10$, the initial configuration is:

$$0, \quad 0, \quad 0, \quad 0, \quad 0, \quad 0, \quad 0, \quad 0, \quad 0, \quad 0$$

We now choose one die at random, throw it and return it to its place. We can think of a machine that scans the sequence of dice, chooses one die at random and gives it a kick so that the next outcome of that particular die will be either "0" or "1" with equal probability. Alternatively, we can play the game on a computer.[2]

With these simple rules, we will follow the evolution of the configurations. But again, if you feel more comfortable, let us play the game. You choose a number between zero and 10, say 4, and I choose a number between zero and 10, say 6. We start with an all-zeros configuration and proceed with the game according to the rules described above. At each step, we check the sum. If the sum is 4, you win a point; if it is 6, I win a point. What sum will you choose? Do not rush to choose the *sum* = 0. You can argue that you *know* the initial configuration, therefore, you will

---

[1]Later, we shall start with any arbitrary configuration. For the beginning, we shall assume that the initial configuration is the "all-zeros" configuration.

[2]The program is very simple. First, it chooses a number from 1 to N, then it changes the face on the die at that particular location, to get a new outcome between 1 and 6.

win with certainty on the zeroth step. Indeed, you are perfectly right. You *will* win on the zeroth step! But what if we decide from the outset to play a million steps?

Let us now examine carefully and patiently how this game will evolve after many steps. You should follow the evolution of the game even if you are interested only in finding the winning number. Following the evolution of the game is crucial to an understanding of the Second Law of Thermodynamics. So be attentive, alert and concentrate on *what* goes on, *why* it goes this or that way, and as for the *how* it goes on, we have already set-up the "mechanism" as described above.

Remember that we are playing a new game. The dice have only two outcomes "0" and "1." Therefore, the sum of the outcomes of N dice can be one of the numbers from zero (all "0") to N (all "1"). Furthermore, there is only one partition for each sum, and this is a great simplification compared with the previous game, where we had to count different weights for the different partitions, a very complicated task for a large N (see Chapters 2 and 3). Here, we have to worry only about the weight of one partition. For example, with $N = 4$ and a choice of $sum = 2$, there is only one partition of 2 in terms of two "zeros" and two "ones." These are

$$0011 \quad 0101 \quad 0110 \quad 1001 \quad 1010 \quad 1100$$

There are six specific configurations, i.e., six ordered sequence of zeros and ones, for the dim event $sum = 2$. By dim event, we mean the number of "ones" that are in the event, or in the configuration, regardless of the specific locations of these "ones" in the sequence. For any chosen dim event, say $sum = n$, there are different ways of sequencing the "ones" and "zeros" that determine a specific configuration, or a specific event.

Before we proceed with the new games, I would like to draw your attention again to Figs. (3.5) and (3.6) (or (3.7) and (3.8)). In these figures we show the probabilities as a function of the various sums, and as a function of the *reduced* sum (i.e., the sum divided by the maximum sum, which is simply the number N). Note that all these plots have a maximum at $N/2$ (or at reduced *sum* = $1/2$). The larger N is, the lower the value of the maximum.

4.1. Two Dice; N = 2

As in the previous case, the game with one die is uninteresting and we shall start by analyzing the case of two dice.

Recall that in the present game, we start with a specific initial configuration, which for the next few games will always be the all-zeros configuration. Suppose you chose the *sum* = 0, arguing that since you *know* the result of the zeroth step, which is *sum* = 0, you are guaranteed to win the zeroth step. Indeed, you are right. What should I choose? Suppose I chose *sum* = 2, the largest possible sum in this game. Recall that in the previous game with two *real* dice, the minimum *sum* = 2, and the maximum *sum* = 12, had the same probability $1/36$. Here, the rules of the game are different. If you choose *sum* = 0, and I choose *sum* = 2, you win with probability one and I win with probability zero on the zeroth step, so you do better on the zeroth step. What about the first step? You will win on the first step if the die, chosen at random and tossed, has the outcome of zero. This occurs with probability $1/2$. What about me? I choose *sum* = 2. There is no way to get *sum* = 2 on the first step. On the first step, there are only two possible sums, zero or one. Therefore, the probability of my winning on the first step is zero too. So you did better both on the zeroth and first steps.

What about the next step? It is easy to see that your chances of winning are better on the second step as well. In order for me to win, the sum must increase from zero to one on the first step, *and* from one to two on the second step. You have more *paths* to get to *sum* = 0 on the second step; the *sum* = 0 can be realized by, "stay at *sum* = 0" on the first step, *and* "stay

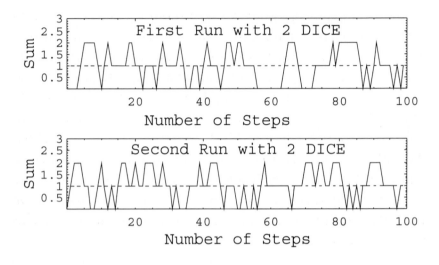

Fig. (4.2)

at *sum* = 0" on the second step, or increase to *sum* = 1 on the first *step*, and decrease to *sum* = 0 on the second step.

Figure (4.2) shows two runs of this game,[3] each run involving 100 steps. It is clear that after many steps, the number of "visits" to *sum* = 0 and to *sum* = 2 will be about equal. Although we started with *sum* = 0, we say that after many steps, the game loses its "memory" of the initial steps. The net result is that you will do a little better in this game.

What if you choose *sum* = 0, and I choose *sum* = 1? In this case, you will win with certainty on the zeroth step. On the second step, you have a probability, 1/2, of winning, and I have a probability, 1/2, of winning. But as we observe from the "evolution" of the game, after many steps, the game will visit *sum* = 0 far less frequently than *sum* = 1. Therefore, it is clear that after many games, I will be the winner in spite of your *guaranteed* winning on the zeroth step.

---

[3]We refer to a "run," as the entire game consisting a predetermined number of "steps."

Let us leave the game for a while and focus on the evolution of the game as shown in the two runs in Fig. (4.2). First, note that initially we always start with *sum* = 0, i.e., with configuration {0, 0}. We see that in one of the runs, the first step stays at *sum* = 0, and in the second run, the sum increases from zero to one. In the long run, we shall be visiting *sum* = 0 about 25% of the steps; *sum* = 2, about 25% of the steps; and *sum* = 1, about 50% of the steps. The reason is exactly the same as in the case of playing the two-dice game in the previous chapter. There is *one* specific configuration for *sum* = 0, *one* specific configuration for *sum* = 2, but *two* specific configurations for *sum* = 1. This is summarized in the table below.

Configuration	{0, 0}	{1, 0} {0, 1}	{1, 1}
Weight	1	2	1
Probability	$^1/_4$	$^2/_4$	$^1/_4$

This is easily understood and easily checked by either experimenting with dice (or coins), or by simulating the game on a computer.

In Fig. (4.2), you can see and count the number of visits at each level. You see that the slight advantage of the choice of *sum* = 0 will dissipate in the long run.

Before we proceed to the next game with four dice $N = 4$, we note that nothing we have learned in this and the next game seems relevant to the Second Law. We presented it here, mainly to train ourselves in analyzing (non-mathematically) the evolution of the game, and to prepare ourselves to see how and why new features appear when the number of dice becomes large. These new features are not only relevant, but are also the very essence of the Second Law of Thermodynamics.

If you run many games like this on a computer, you might encounter some "structures" in some of the runs, for instance, a sequence of 10 consecutive zeros, or a series of alternating zeros and ones, or whatever specific structure that you can imagine. Each specific "structure", i.e., a specific sequence of results *is* possible and if you run many games, they will occur sometimes. For instance, the probability of observing the event *sum* = 0, in all the 100 steps is simply $(1/2)^{100}$ or about one in 10^{30} steps.

4.2. Four Dice; N = 4

We proceed to the game with four dice. This introduces us to a little new feature. Again, we start the game with an all-zeros configuration at the zeroth step, pick-up a die at random, throw it, and place the die with the new face in its original place on the line. Figure (4.3) shows two runs of this kind. Each game is run with 100 steps. We plot the *sum* as a function of the number of steps. The sum ranges from the minimum *sum* = 0 (all-zeros), to the maximum *sum* = 4 (all-ones).

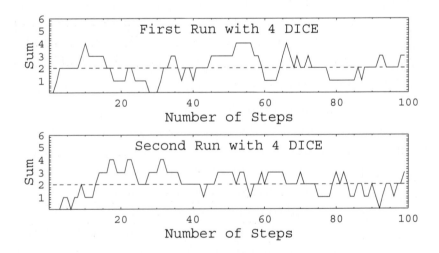

Fig. (4.3)

If you want to play by choosing *sum* = 0, and stick to this choice (as required by the rules of the game), you will win if I choose 4. The reason is the same as in the previous two-dice game. Since we started with *sum* = 0, the game is slightly "biased" towards *sum* = 0. If I choose *sum* = 2, I will certainly lose on the zeroth step and you will win with certainty the zeroth step. You will also have some advantage in the next few steps. However, your initial advantage will dissipate in the long run. You can see that within 100 steps, we have visited on the average:

$$Sum = 0 \quad \text{in} \quad \tfrac{1}{16} \text{ of the steps}$$
$$Sum = 1 \quad \text{in} \quad \tfrac{4}{16} \text{ of the steps}$$
$$Sum = 2 \quad \text{in} \quad \tfrac{6}{16} \text{ of the steps}$$
$$Sum = 3 \quad \text{in} \quad \tfrac{4}{16} \text{ of the steps}$$
$$Sum = 4 \quad \text{in} \quad \tfrac{1}{16} \text{ of the steps}$$

These average numbers are for the long run, and can be calculated exactly. As you can see, the slight advantage of choosing *sum* = 0 will dissipate in the long run. Let us examine more carefully the runs of this game (Fig. (4.3)) and compare it with the two-dice game (Fig. (4.2)).

An obvious feature of this game is that the total number of visits to the initial configuration (*sum* = 0), is much smaller than in the previous game. Let us examine all the possible configurations of the system. Here, they are:

Dim Event	Specific Events
sum = 0	0,0,0,0
sum = 1	1,0,0,0 0,1,0,0 0,0,1,0 0,0,0,1
sum = 2	1,1,0,0 1,0,1,0 1,0,0,1 0,1,1,1 0,1,0,1 0,0,1,1
sum = 3	1,1,1,0 1,1,0,1 1,0,1,1 0,1,1,1
sum = 4	1,1,1,1

Altogether, there are 16 possible *specific* configurations. We have grouped these configurations into five groups; each of these is characterized by only the *sum* of the outcomes, or equivalently, by the "number of ones," or the number of dots, irrespective of where the "ones" appear in the sequence. Thus, we have proceeded from 16 *specific* configurations (i.e., specifying exactly the locations of all the zeros and ones), to *five dim* configurations (where we specify only the number of "ones'). The distinction between the specific and dim configurations is very important.[4] The dim configuration always consists of one or more specific configurations.

Another feature which is common to both games (and to all the remaining games in this chapter) is that though there is initially a slight bias to the visits *sum* = 0 compared with *sum* = 4, in the long run (or very long run), we shall see that on the average, the sum will fluctuate above and below the dotted line we have drawn at the level of *sum* = 2. This feature will become more pronounced and clearer as N increases.

Before moving on to the next game, you should "train" yourself with this game either by actually playing dice, or by simulating the game on a PC. You should only use your common sense to understand why the number of visits to the initial state decreases when N increases, and why the number of visits to *sum* = 2 is the largest in the long run.

4.3. Ten Dice; $N = 10$

With 10 dice, we shall observe something new, something that will come into greater focus as we increase the number of dice, until we can identify a behavior similar to the Second Law.

---

[4]In the molecular theory of the Second Law, the specific and the dim configurations correspond to the microstates and macrostates of the system. We shall discuss this in Chapter 7.

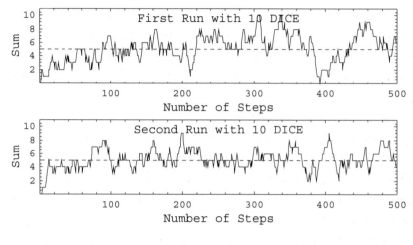

Fig. (4.4)

We start as before, with an all-zeros configuration, i.e., *sum* = 0, i.e., the number of "ones" is zero. Figure (4.4) shows two runs of this game.

At the first step we have the same probability of staying at *sum* = 0, or move to *sum* = 1; these have an equal probability of 1/2. We could barely see that on the scale of Fig. (4.4). The important thing that is happening is on the second, third and perhaps up to the 10 next steps. As we can see in the runs shown in Fig. (4.4), the overall trend in the first few steps is to *go upwards*. Why? The answer is very simple. After the first step, we have one of the two configurations; either all-zeros, or nine zeros. We now choose a die at random. Clearly, it is much more likely to pick up a zero, then a one. Once we pick up a zero die, we can either go upwards or stay at the same level, but not downwards. To go downwards we need to pick up a "1" (with relatively low probability) and turn it into "0" with probability 1/2. So, going downwards on the second step becomes a rare event (and it will be more and more so as N increases). I urge you to examine carefully the argument given above, and convince yourself that on the second step, it is far more likely to go upwards than downwards. It is important to fully understand this behavior at this stage, before we go on to a larger number of dice. The reason I urge you to do the calculations for 10 dice, then extrapolate for a larger number of dice, is because with 10 dice, the calculations are very simple. For a larger number of dice, the calculations might be intimidating and thus, discourage you from analyzing all the probabilities.

Here, it is very simple. We start with an all-zeros configuration. Therefore, at the first step, we choose a "zero" with probability one, then we can either go upwards with probability $1/2$, or stay at the same level (*sum* $= 0$) with probability $1/2$. There is no going downwards.

Similarly, on the second step, the argument is a little more complicated. If we end up on the first step at the level *sum* $= 0$, then the two possibilities are exactly as for the first step. However, if we end up at the first step at level *sum* $= 1$, then we have four possibilities:

1) Pick up at random a "1" with probability $1/10$, *and* stay at the same level with probability $1/2$
2) Pick up at random a "1" with probability $1/10$ *and* go downwards with probability $1/2$
3) Pick up at random a "0" with probability $9/10$ *and* go upwards with probability $1/2$
4) Pick up at random a "0" with probability $9/10$ and stay at the same level with probability $1/2$.

The net probabilities of the four possibilities are:

1) $1/10$ times $1/2 = 1/20$ for staying at the same level (*sum* $= 1$)
2) $1/10$ times $1/2 = 1/20$ for going downwards (*sum* $= 0$)
3) $9/10$ times $1/2 = 9/20$ for going upwards (*sum* $= 2$)
4) $9/10$ times $1/2 = 9/20$ for staying at the same level (*sum* $= 1$).

Clearly, staying at the same level (*sum* = 1), or going upwards has a much higher probability than going downwards. This is reflected in the general upward trend in the runs shown in Fig. (4.4).

On the third step, we can do the same calculation. The trend to go upwards is still greater than to go downwards, although a little weaker as for the second step. Why? Because to go upwards, we need to pick up a "0" with probability of at most $8/10$ (presuming we are at the level *sum* = 2 at this step) and climb up with probability $1/2$. The probability of climbing up is still larger than that of going down, but it is somewhat weaker compared with the upward trend at the second step. The argument is the same for the fourth, fifth steps, etc.; with each increasing step, the probability of climbing up becomes less and less pronounced until we reach the level *sum* = 5. When we reach this level (*sum* = 5), again we have four possibilities:

1) Pick up a "1" with probability $1/2$, *and* stay at the same level with probability $1/2$
2) Pick up a "1" with probability $1/2$, *and* go downwards with probability $1/2$
3) Pick up a "0" with probability $1/2$, *and* go upwards with probability $1/2$
4) Pick up a "0" with probability $1/2$, *and* stay at the same level with probability $1/2$.

The net probabilities of these four possibilities are:

1) $1/2$ times $1/2 = 1/4$ for staying at the same level (*sum* = 5)
2) $1/2$ times $1/2 = 1/4$ for going downwards (*sum* = 4)
3) $1/2$ times $1/2 = 1/4$ for going upwards (*sum* = 6)
4) $1/2$ times $1/2 = 1/4$ for staying at the same level (*sum* = 5).

The important finding is that once we reach the level *sum* = 5, we have probability $1/4$ of going upwards and the same probability, $1/4$ of going downwards, but twice that probability, i.e. $1/2$ of staying at that level, *sum* = 5.

Once we reach the level *sum* = 5 for the first time, the chances of going upwards or downwards become symmetrical. We may say that at this stage, the system has "forgotten" its initial configuration. If we start with any arbitrary configuration at a level below *sum* = 5, there will be a greater tendency to go upwards than to go downwards. If on the other hand, we start at a level *above sum* = 5, we shall have a strong bias to go downwards. If we start at level *sum* = 5, or reach that level during the run, we shall have a larger probability of staying at that same level. As we shall see below, these arguments are valid for all the cases, but become more powerful for a larger number of dice. For this reason, we shall refer to the level *sum* = 5 (in this particular game with 10 dice) as the *equilibrium* level. Indeed, the very nature of this level is to be at *equilibrium*, i.e., with equal *weights*, or equal probabilities of going upwards or downwards but a larger probability of staying put. We drew the equilibrium line with a dashed line in each run. It is an equilibrium also in the sense that for any deviation from that line, either upwards or downwards, the system has a tendency of returning to this line. We can think of an imaginary "force" restoring the system back to the equilibrium level. The further we go away from the equilibrium line, the larger will be the restoring "force" leading us back to the equilibrium line.

The two features described above — initially a preference to go upwards, and once reaching the equilibrium level, to stay there or around there — are the seeds of the Second Law. Figuratively, it can be described as a kind of ghostly "force" that attracts any specific configuration towards the equilibrium line. Once there, any deviation from the equilibrum line is "pulled" back to that line.[5] To put it differently, we can think of the equilibrium line as an *attractor*, always pulling the "sum" towards it.

For this reason, large deviations from the equilibrium line have very little chance of occurrence. This is why we have observed only very rarely the reverting of a configuration to the original state, or a configration reaching the extreme level of *sum* $= 10$. (It is easy to calculate that a visit to either of these extremes has a probability of $(1/2)^{10}$, which is about one in 1000 steps.)

As with the four dice, we could have listed all the possible specific configurations for the 10 dice. In the table below we have listed only a few configurations in each group of the dim-state.

Dim Event	Examples of Specific Events	Number of Specific Events Comprising the Dim Event
dim-0	0000000000	1
dim-1	0000000001, 0000000010,...	10
dim-2	0000000011, 0000001010,...	45
dim-3	0000000111, 0000001011,...	120
dim-4	0000001111, 0000010111,...	210
dim-5	0000011111, 0000101111,...	252
⋮	⋮	⋮
dim-10	1111111111	1

---

[5]Note, however, that the equilibrium line does not characterize a *specific* configuration. In fact, this is a dim configuration consisting of the largest number of specific configurations for this particular game. The meaning of the equilibrium line here is related to, but not the same, as the equilibrium *state* of a thermodynamic system, as discussed in Chapter 7.

We have grouped all the $2^{10} = 1024$ *specific* configurations into 11 groups; each group is referred to as a *dim* configuration (or dim-state or dim-event). Dim-one (i.e., all configurations having only a single "1") has 10 specific configurations. Dim-five has 252 specific configurations.

Although I did not recommend pursuing the game aspect of this experiment, you might have taken note that if you have chosen *sum* = 0, and I have chosen *sum* = 5, initially you will win at the zeroth step with certainty. You will win on the next step with probably $1/2$, and thereafter the probability will decline rapidly as the game proceeds. On the other hand, I have zero probability of winning in the first few steps. However, once we reach the level *sum* = 5, I will have the upper hand, and on the average, the odds of winning are 252:1 in my favor!

Let us turn to examine the evolution of the next two runs before we unfold the full characteristic of the behavior of what entropy essentially is. Let us also change our nomenclature. Instead of saying level *sum* = *k*, we shall simply say dim-*k*. The *sum* is really not the important thing. What is of importance is how many "ones," or how many dots there are in the dim configuration, or the dim event.

4.4. Hundred Dice; N = 100

The runs shown in Fig. (4.5) were taken with 1000 steps. Evidently, with a larger number of dice, we need more steps to go from dim-0 to the equilibrium level, which in this case is dim-50. How fast do we get to the equilibrium level? If we are extremely lucky, and on each step we pick up a zero, and on each throw we get a "1," then we need a minimum of 50 steps to reach the level dim-50 for the first time.[6] However, we do not pick up a "0"

---

[6]In the real process, it is mainly the temperature that determines the rate of the process. However, neither in this game nor in the real process shall we be interested in the *rate* of the process (see also Chapter 7).

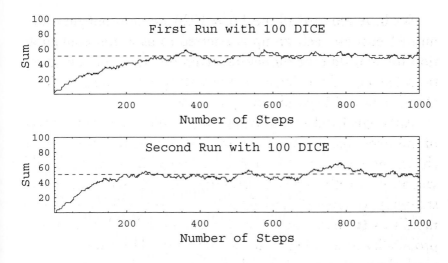

Fig. (4.5)

on each step, and when we pick up a zero, it has only a chance of $1/2$ to convert to "1," i.e., going upwards. These affect the rate of climbing towards the level dim-50. You can see that on the average, we get there in about 200 to 400 steps. As you can observe, the ascension is quite steady, with occasionally resting at the same level, and occasionally sliding down for a while, and then regaining the upward thrust.

Once you get to level dim-50, you will be staying at that level or in its vicinity, most of the time. Occasionally, a large deviation from the equilibrium level will occur. During these specific runs, the initial level dim-0 was never visited. This does not imply *never* visiting level zero. The chance of this occurrence is one in $(2)^{100}$ steps or one in about 10^{30} = $1,000,000,000,000,000,000,000,000,000,000$ steps. (This is a huge number [just "huge," and we can even write it explicitly. However, it is quite small compared to what awaits us in the next runs: numbers that we could never even write explicitly] the chances of visiting level dim-0 (or level dim-100) is less than once in a billion of a billion of steps.) Do not try to do this experiment with dice. This is a long, arduous and tedious experiment. It is easier to run the game on a PC.

4.5. Thousand Dice; N = 1000

The results of this run are much the same as before, but having a slightly smoother curve, on the scale of Fig. (4.6). To reach the equilibrium level, we need about 3000 to 4000 steps. Once we reach the equilibrium line, deviations from this level have about the width of the line demarcating the equilibrium level. Larger deviations are rare (can be seen as sharp spikes like barbed wire stretched along the equilibrium line), and of course, deviations large enough to bring us to level zero will "never" be observed. This is not an impossible event but it will occur at a rate of about one in 10^{300} steps (one followed by three hundred zeros). Even if we run the program for a very long time on the computer, we could still not observe this event.

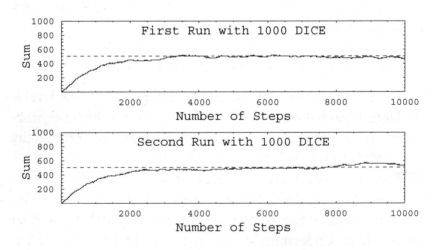

Fig. (4.6)

Recall our calculations of the probability of going upwards, going downwards, or staying at the same level. Although the calculations are the same, it is instructive to have a feel for how these tendencies change with the number of dice.

On the first step, we have the same probability of going upwards or of staying at level dim-0 as before. Suppose we reached dim-1 after the first step; then we have four possibilities for the second step:

1) Pick up at random a "1" with probability $1/1000$ *and* stay at that level with probability $1/2$
2) Pick up at random a "1" with probability $1/1000$ *and* go downwards with probability $1/2$
3) Pick up at random a "0" with probability $999/1000$ *and* go downwards with probability $1/2$
4) Pick up at random a "0" with probability $999/1000$ *and* stay at the same level with probability $1/2$.

Thus, the probabilities of the three *net* possibilities for the next steps are:

1) $1/1000$ times $1/2 + 999/1000$ times $1/2 = 1/2$ to stay at the same level
2) $999/1000$ times $1/2 = 999/2000$ for going upwards
3) $1/1000$ times $1/2 = 1/2000$ for going downwards.

We note that the probability of staying at the same level is the largest (probability nearly $1/2$). This is almost the same probability as for going upwards , 999/2000, while the probability of going downwards is negligible, ($1/2000$).

This is a very remarkable behavior, and you should examine these numbers carefully. Compare this with the case of $N = 10$, and think of what happens when N increases to 10^4, 10^5 and far beyond that. Understanding this game is crucial to the understanding of the way entropy behaves.

The same calculations can be repeated for the third, fourth, etc., steps, As long as there are many more zeros in the configurations, there will be a larger probability of going upwards and a smaller probability of going downwards. The "strength" of this argument becomes weaker and weaker as we climb up and as the sum reaches the level $N/2 = 500$. This is reflected in the form of the overall curve we observed in Fig. (4.6). Initially the climb upwards has a steep slope; then the slope becomes more and more gentle as we get closer to the equilibrium line. Once we reach that level for the first time, we have a larger probability of staying there, and equal probabilities of going either upwards or downwards. As in the previous cases, whenever any deviations occur from the equilibrium line, the system will have a tendency to return to this line, as if there is an unseen "force" pulling the curve towards the equilibrium line.

4.6. Ten Thousand Dice; $N = 10^4$ and Beyond

Figure (4.7) shows a run with $N = 10^4$. There is no need to draw more than one run. All the runs are nearly identical in shape. As we can see, the curve is very smooth on this scale; even the little spikes we have observed in the previous case have disappeared. This does not mean that up and down fluctuations do not occur; it only means that on the scale of this graph, these fluctuations are unnoticeable. We can notice these fluctuations if we amplify the curve at some small range of steps as we have done in the two lower panels of Fig. (4.7).

As you can see, once we have reached the equilibrium line (this occurs after quite a large number of steps), we stay there and in its vicinity, almost permanently. The curve of the results of our run, and the equilibrium line merges and becomes one line. We do not see any major fluctuations and certainly there

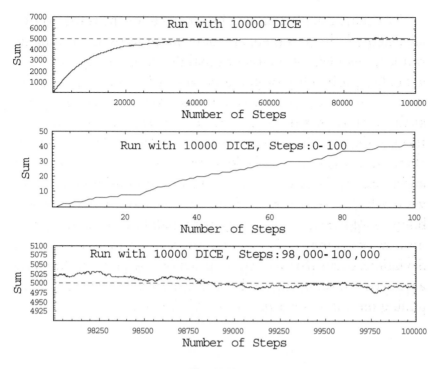

Fig. (4.7)

is no visit to the initial configuration. However, there is still a non-zero probability of occurrence of such an event. This is $(1/2)^{10,000}$ or about once in every 10^{3000} steps (one followed by 3000 zeros; do not try to write it explicitly). This means that practically, we shall "never" see any visits to the initial configuration and once we reach the equilibrium line, we will stay near that level "forever." I have enclosed the words "never" and "forever" within quotation marks to remind you that neither "never" nor "forever" are *absolute*, i.e., "once in a while" *there is* a non-zero chance of visiting the initial configuration. This chance is already extremely small for $N = 1000$, and as we shall see in Chapter 7 (when dealing with real systems), we have numbers of the order of $N = 10^{23}$, which is billions and billions times larger than $N = 1000$. For such a system, the probability of visiting the starting configuration is so small that we can actually use the words "never" and "forever" without quotation marks.

To give you a sense of what this number means, think of running the game at a rate of 1000 steps in a second. If you think you can do it faster, do it with one million steps a second. The age of the universe is currently estimated to be on the order of 15 billion years. Therefore, if you were playing this game with one million steps per second, you will do the total of

$$10^6 \times 60 \times 60 \times 24 \times 365 \times 15 \times 10^9 = 4 \times 10^{16} \text{ steps}$$

That is, you will be doing about 10,000,000,000,000,000 steps during all this time. This means that if you play the game during the entire age of the universe, you will not visit the initial configuration even once. You will have to play a billion times the *age* of the universe in order to visit the initial configuration. Thus, although we have admitted that "never" is not *absolute*, it is very close to being absolute. Let us ponder for a while what is really meant by an absolute "never" or "forever."

In daily life you might use these terms without giving it a second thought. When you promise your wife to be faithful "forever," you don't mean it in the absolute sense. The most you can say is that within the next hundred years you will "never" betray her.

What about the statement that the sun will rise every morning "forever?" Are you sure that this will occur "forever" in an absolute sense? All we know is that in the past few million years, it did occur, and we can predict (or rather guess) that it will continue to do so in the next millions or billions of years. But can we say "forever" in an absolute sense? Certainly not!

What about the laws of nature? Can we say that the law of gravity, or the speed of light, will remain the same "forever?"

Again, we are not sure;[7] all we can say is that it is probable that they will hold out at most, for the next 15 billion years or so.

Thus, there is no way of saying that any event in the real physical world will occur "forever," or will "never" occur in an absolute sense. Perhaps, these terms can be applied to the Platonic world of *ideas*. In that world, we can make many statements like "the ratio of the circumference to the radius of a circle" will *never* deviate from 2π.

We have enclosed the words "never" and "forever" within quotation marks to remind us that these are not meant to be in an absolute sense. We have also seen that in the physical world, we can *never* be sure that we can use "never" and "forever" in an absolute sense (the first *"never"* in the preceding sentence is closer to the absolute sense, than the second "never"). But we have seen that the assertion made about staying *"forever"* at, or near the equilibrium line, and *"never"* returning to the initial state, can be used with much more confidence than in any other statement regarding physical events. In the context of the Second Law, we can claim that "never" is the "neverest" of all "nevers" that we can imagine, and the staying "forever" at equilibrium is the "foreverest" of all "forevers."

Therefore, from now on, we shall keep the quotation marks on **"never"** and **"forever,"** to remind us that these are not absolute, but we shall also render these words in bold letters to emphasize that these terms, when used in the context of the Second Law, are closer to the absolute sense than any other statements made in the real physical world.

I hope you have followed me so far. If you have not, please go back and examine the arguments for the small number of

---

[7]Recently, even the constancy of the constants of physics has been questioned. We do not really know whether or not these constants, such as the speed of light did not, or will not change on a scale of cosmological times. See Barrow and Webb (2005).

dice. In all of the games in this chapter you should ask yourself three questions:

1) *What* is this *thing* that changes at each step? Try to give it a name of your choice. A name that reflects what you were monitoring in any of these games.
2) *How* is this change achieved? This is the easiest question! And most importantly,
3) *Why* is the change always in one direction towards the equilibrium line, and why it "**never**" goes back to the initial configuration, and stays very near the equilibrium line "**forever**."

If you feel comfortable with these questions and feel that you can explain the answer, you are almost 50% towards understanding the Second Law of Thermodynamics.

We still have one more important question to answer. What have all these dice games to do with a real physical experiment, and to what extent do the things that we have observed have relevance to the Second Law? We shall come back to these questions in Chapter 7. But before doing that, we shall take a break.

I presume it took some effort on your part to follow me thus far. Therefore, I suggest that it is time to relax and experience, somewhat passively, the Second Law (or rather, some analogues of the Second Law) with our senses. What we will do in the next chapter is simply to imagine different experiments which, in principle, you could experience with your senses. This should give you a sense of the variety of processes that can actually occur under the same rules of the game. This will also prepare you to better appreciate the immensely larger number of processes that are actually governed by the Second Law. You need not make any calculations; just think of why the processes you are experiencing occur in this or that particular way. We shall come back to a deeper analysis after that experience and also translate all that we have experienced from the dice language into the language of real systems consisting of atoms and molecules. In Chapter 6, we shall try to understand the Second Law within the dice-world. In Chapter 7, we shall translate all this understanding into the language of real experiments.

Experience the Second Law with all Your Five Senses

In this chapter, we shall not gain any new understanding of the games of dice as we have done in the previous chapters, nor obtain any new insights into the nature of the Second Law of Thermodynamics. Instead, we shall describe a few hypothetical experiments. Underlying all of these processes is a common basic principle which, as we shall see, *is* relevant to the Second Law. We shall also use different senses to *feel* the Second Law. All of these processes are different manifestations of the same principle and represent the immense number of possible manifestations of the Second Law. To avoid boring you with repetitions, we shall change the rules of the game slightly. This is done to demonstrate that the rules need not be rigid. For instance, we do not have to pick up a single die in each step; we can pick up two dice in each step, or three or four, or even all of them at once. We can also pick up the dice in an orderly fashion, not necessarily at random, but change the outcome at random; or we can choose a die at random but change the outcome deterministically, e.g., from "1" to "0," or from "0" to "1." What matters is that each die has a "fair" opportunity to change, and that there is an element of randomness in the process. We shall return to a discussion of the mechanism of the change in Chapter 7. But for now, let us enjoy some hypothetical experiments. They are designed to be enjoyable and to enrich your experience and familiarity with the

Second Law. You do not need to make any effort to understand these processes. We shall come back to the issue of *understanding* in the next chapter, and to the issue of *relevance* to the real world in Chapter 7. For now, just read and enjoy the "experience."

5.1. See it with your Visual Sense

We start with the simplest case. We do not change any of the rules as described in Chapter 4, nor do we change the number of possible outcomes, nor their probabilities. The only difference is that instead of *counting* the number of "ones" (or equivalently summing all the outcomes of all the dice), we watch how the *color* of a system changes with time, or with the number of steps.

Fig. (5.1)

Suppose we have N dice, three faces of each die being colored blue and the other three, yellow. You can think of a coin being on one side blue and yellow on the other, but we shall continue to use the language of dice in this and the next chapters.

A specific configuration of the N dice will be a specification of the exact color shown on the upper face of each die. In this particular experiment, there is no "sum" of outcomes that we can count. The outcomes are colors and not numbers.[1] In the

---

[1]In physical terms, the outcome is an electromagnetic wave of a certain frequency. This specific wave enters the eye, and focuses on the retina's rod cells which send a message to the brain. There, it is processed and we perceive that signal as a *color*.

case of numbered dice, we have referred to a dim configuration, the set of all the specific configurations that have the same sum or equivalently the same number of "ones." To proceed from the *specific* configuration to the *dim* configuration, you can think of the dice as pixels in a digital picture; pixels of two types of colors so small that you observe only an "average" color. In Fig. (5.1), we show a few dim events. On the extreme left bar, we have (100%) yellow. We add (10%) of blue to each successive bar. The bar at the extreme right is pure (100%) blue.

In all of the examples in this chapter, we use $N = 100$ dice. Initially, we start with an all-yellow dice and turn on the mechanism of changing the outcomes of the dice with the same rules as before. Choose a die at random, throw it and place it back into the pool of dice. These are admittedly very artificial rules. However, we shall see in Chapter 7 that, in principle, such an evolution of colors can be achieved in a real physical system where the "driving force" is the Second Law of Thermodynamics.

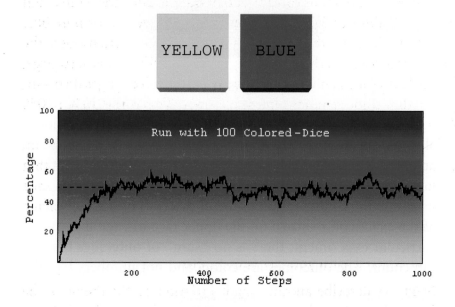

Fig. (5.2)

What are we going to observe? Of course, here we cannot "plot" the sums of the outcomes. We could have assigned the number "zero" to the yellow face, and the number "1" to the blue face, and plot the number of ones as the game evolves with time or with the number of steps. But in this experiment, we only want to *see* what goes on. In Fig. (5.2), we show how the "average" color evolves with time. Starting with an all-yellow configuration, the ascension is measured as the *percentage* of the added blue color.

If we start the run with an all-yellow system, we will initially observe nothing. We know that there is a strong "upward" tendency towards blue as we have observed in Chapter 4. The effect of a few blue dice will not be perceived by our visual sense. As we proceed with many more steps, we will start to see how the system's color slowly changes from yellow to green. Once we reach a certain hue of green (the one which is composed of about 50% blue and 50% yellow), the color of the system remains permanent. We will never observe any further changes. Perhaps there will be some fluctuations about the equilibrium "green line," but we will almost never "visit" either the pure yellow or pure blue colors. Note also that even for $N = 100$, all the fluctuations in the colors are within the range of green color. When N is very large, no fluctuations will be observed, and although changes do occur, the dim color remains almost constant. This is the 50% : 50% mixture of the blue and yellow colors.

We will show in Chapter 7 that such an experiment can actually be carried out with real particles (say, two isomers that have different colors). For the moment, we should only take note that from whatever configuration we start with, applying the rules will lead us to the same final (equilibrium) color.

5.2. Smell it with your Olfactory Sense

Next, we describe another small variation on the theme of the Second Law. As in the previous example, we have again two kinds of faces with equal probabilities. We assume that when a die shows a face A, it emits a scent of type A, and when it shows a face B, it emits a scent of type B. If we have a hundred dice, $N = 100$, with any arbitrary configuration of As and Bs, we smell an average scent of type $A + B$, in the same proportion as the ratio of As to Bs in that configuration. Note that the sense of smell is a result of specific molecules adsorbed on specific receptors.[2]

As we shall discuss in Chapter 7, an analogue of this experiment can be performed by using real molecules. In principle, we can actually follow the evolution of the scent of a system with our olfactory sense, as it evolves according to the Second Law of Thermodynamics. Here, however, we discuss only a toy experiment with dice that emit two different molecules having different scents.

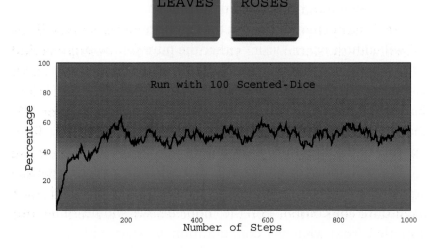

Fig. (5.3)

---

[2]The physical outcomes are molecules of specific structures that are adsorbed on receptors that are located on the surface of the inner part of the nose. From there, a signal is transmitted to the brain where it is processed to produce the perception of a specific scent.

Again, we start with an all-*A*s configuration and start running the game as in Chapter 4. However, there is a slight variation in the rules. We pick up *two* dice at random, throw them and smell the resulting configuration. The throwing process produces either *A* or *B*, with probabilities 1/2 and 1/2. For example, suppose that face *A* emits molecules that have the smell of green leaves, and face *B* emits molecules that have the scent of red roses (Fig. (5.3)).

Initially, we smell the pure scent of type *A*. After several steps, we still smell type *A* although we know that there is a high probability of the system going "upwards," i.e., towards a mixture of *A* and *B*. However, a small percentage of *B* will not be noticeable even by those trained professionally in sniffing perfumes (e.g. people who are employed in the cosmetic and perfumery business). In Fig. (5.3), we show the evolution of the scent of the system with 100 dice. We start with the leafy scent of green leaves and run the game.

After many steps, you will start to notice that scent type *B* has mixed with the dominating scent type *A*. After a longer period of time, you will reach a point where the ratio of the scents *A* to *B* will be 1:1, i.e., in accordance with the "equilibrium line" of the configurations of the dice having the ratio of 1:1. Once we have reached that specific blend of scent, we will no longer experience any further changes. We know that there are some fluctuations, but these will not be very different from the equilibrium scent. As the number of dice increases, we will reach a constant equilibrium scent; no noticeable deviations from this scent will be sensed.

As we have noted above, although this is a very hypothetical process, one can actually design a *real* experiment and follow the evolution of the overall scent of a mixture of molecules having different scents. We will discuss this kind of experiment in Chapter 7. In fact, the real experiment is much easier to design than this one with hypothetical dice.

5.3. Taste it with your Gustatory Sense

As with the sniffing experiment, we can design an experiment in which we can taste with our tongue. In this experiment, we change one aspect of the rules of the game. Again, we start with 100 dice. Each die has three faces having a *sweet* taste, say sugar syrup, and the other three having a *sour* taste, say lemon juice (Fig. (5.4)). We proceed to find out the "average" taste of the entire system.[3] The taste that we experience is a dim taste. We do not distinguish between different specific configurations of the dice; only the ratio of the two tastes can be perceived by our gustatory sense.

We begin with an all-sour configuration (represented by yellow in Fig. (5.4)). Instead of choosing a die at random, we

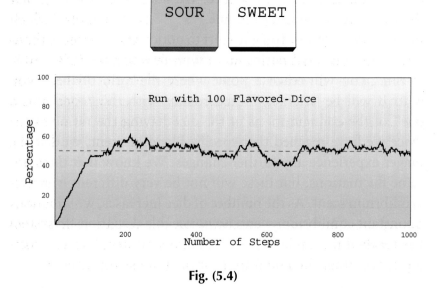

Fig. (5.4)

---

[3]As in the case of scents, the sense of taste is produced by molecules of a specific structure that are adsorbed on sensitive taste cells in microscopic buds on the tongue. From there, a message is sent to the brain where it is processed to produce the sensation of taste.

choose a die in an orderly fashion, say sequentially from left to right. We pick up a die, throw it, taste the new dim configuration, and then turn to the next die in the sequence, and so on. If we have 100 dice, we will start with the die numbered one, and proceed to the last one, numbered 100. We shall repeat the sequence 10 times, performing 1000 steps in total. Figure (5.4) shows the evolution of the taste in this game.

In spite of the change in the rules of the game, the evolution of the game in its gross features will be the same as before. We start with a pure sour taste and for the first few steps we will not notice any change in the overall taste. Unless you are a gourmet or have an extremely discriminating sense of taste, you cannot tell the difference between a 100% sour taste and a mixture, say of 99% sour and 1% sweet tastes. But we know from our analysis in Chapter 4 (or as you can experiment with dice or carry out a real experiment as described in Chapter 7), that there is a strong tendency for the system to go "upwards," which in this case is towards the sweet-and-sour taste. After a thousand steps, you will notice that the taste is almost 50% : 50% sweet-and-sour, and once you have reached this "level," you will not notice any more changes in the taste. The mechanism of changing the taste of each die is, of course, the same as for the previous examples. However, the changes are such that the dim taste, like the total sum of our dice in Chapter 4, does not change in a noticeable manner. We shall taste the same sweet-and-sour taste "forever." Although we know that fluctuations occur, these are barely noticeable on our tongue. The system has reached the equilibrium sweet-and-sour line, "never" visiting again either the initial sour, or the pure sweet taste.

5.4. Hear it with your Auditory Sense

In this experiment, we describe a hypothetical process that will allow your ears to hear and experience the Second Law. Again

Fig. (5.5)

we make a slight change to the rules; instead of two outcomes as we had in the previous section, we assume that we have three possible outcomes. Again using the language of dice, let us assume that each die has two faces marked with *A*, two marked with *B* and two, *C* (Figure (5.5)). We can imagine that whenever face *A* is shown on the upper side, a tone *A* is emitted. We can think of the faces as vibrating membranes which vibrate at different frequencies. The sound waves emitted are perceived as different tones by our ears.[4] The sound does not have to be produced by the die itself; we can think of a signal given to a tuning fork which produces tone *A* whenever *A* is the outcome of the die. Similarly, another tone, *B*, is produced by outcome *B*, and another tone, *C*, produced by outcome *C*.

Again, we start with an all-*A*s initial configuration and start running the process with the same rules as before, except that we have three, instead of two, "outcomes". Pick-up a die at random and toss it to obtain one of the outcomes, *A*, *B* or *C*.

Using almost the same analysis as we have done in Chapter 4, we can follow the evolution of the system not in terms of a "sum" of outcomes, but in terms of the sound of the dim tone we hear with our ears.

Initially, we will hear a pure *A* tone. We know that there is a high probability of the system "climbing" not to a new "sum,"

---

[4]In physical terms, the sound waves reaching the ear drum produce vibrations that are transmitted to the inner part of the ear. From there, a message is sent to the brain where it is processed to produce the sensation of a specific tone.

but to a mixture of tones, still dominated by A. If we start with $N = 1000$ dice of this kind, we will initially notice almost no change. If you have a good musical ear, after some time you will start hearing a mixture of tones (or a chord) which could be more pleasant to your ears if A, B and C are harmonious.

After a long time we will reach an equilibrium tone. We will hear one harmonious chord composed of the three pure tones A, B and C with equal weights. Once we have reached that "level" of tone, the system will stay there "forever." There will be some fluctuations in the relative weights of the three tones but these will hardly be noticeable even to the most musically trained ears.

5.5. Feel it with your Touch (Tactile) Sense

In this last example, we will describe an extremely hypothetical version of a real experiment involving temperature.

We perceive the temperature of a hot or cold body through our skin.[5] It is a sense for which the molecular origin was not understood for a long time. Nowadays, the molecular theory of heat is well understood; nevertheless, it is still not easy for the layman to accept the fact that temperature is nothing but the "average" speed of motion (translation, rotation and vibration) of the atoms and molecules that constitute matter. We *feel* that a block of iron is cold or hot, but we do not *feel* the motion of the iron atoms. In our daily lives, we regard the two, very different notions, *temperature* and *motion*, as two unrelated phenomena. A fast moving ball could be very *cold*, and an immobile ball could be very *hot*. Yet, one of the great achievements of the molecular theory of matter is the identification of temperature (that we sense as hot and cold) as an average velocity of the

---

[5]The sense of touch is produced by nerve cells underneath the skin which respond to pressure and temperature. The cells send messages to the brain where it is processed to produce the sense of pressure, temperature and perhaps of pain too.

atoms and molecules. This insight was not easy to accept before the establishment of the atomic theory of matter. Nowadays, however, this theory is well established and well accepted.

To capitalize on the insight we have already gained with the dice games, we will design our final process based on our sense of touch. It involves temperature, but temperature in an extremely simplified version of any real experiment. We shall briefly discuss real experiments involving the temperature of gases in Chapter 7. This experiment is very important since the Second Law of Thermodynamics was borne out of the considerations of heat engines and heat flow between bodies at different temperatures.

This experiment is designed specifically to be felt by our fifth and last (recognized) sense. The dice in this game have two kinds of faces, three faces being hot (say 100°C) and three faces being cold (say 0°C).[6] Each face has a fixed temperature.[7] We also assume that the faces are perfectly insulated from each other (otherwise the *real* Second Law of Thermodynamics will be working on the molecules that constitute the dice themselves to equilibrate the temperature within each die, and that will spoil our game). In Fig. (5.6), we show the cold face as blue, and the hot face as red, as commonly used on water taps.

---

[6]This is an extremely hypothetical experiment. In a real experiment that we shall discuss in Chapter 7, atomic particles will replace the dice. While we can imagine molecules having different colors, tastes or smells, there is no "temperature" that can be assigned to each molecule. The temperature we sense is a result of the distribution of kinetic energies among the molecules. To simulate something similar to a real experiment, we should use dice having an infinite number of faces, each representing a different possible velocity of a molecule.

[7]We also need to assume that we have a mechanism that keeps each hot face and each cold face at their fixed temperature. Here, it is quite difficult to prevent the equilibration of temperature between the different faces as well as maintaining a constant temperature of the faces after touching with our hands or with a thermometer.

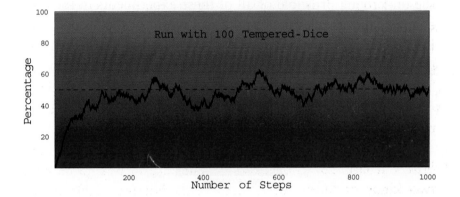

Fig. (5.6)

We start with, say 100 dice, all with the cold faces facing upward, so that if we touch the sample as a whole we feel a cold sensation. The game is run exactly as for the game in Chapter 4, but with a slight change in the rules. We start with all-cold faces upward, selecting a die at *random*. But the change of the face is done not by throwing the die, but deterministically. If it is cold, it will be changed to hot, and if it is hot, it will be changed to cold.

If the dice are very small like "pixels" at different temperatures, we only feel the *average* temperature of the system when we touch the entire system of 100 dice. We cannot discriminate between different *specific* configurations (i.e., which die's face is hot or cold); only the dim configuration, or the dim temperature is felt (i.e., only the ratio of hot to cold dice). As we proceed using these rules, we will feel a gradual increase in the temperature. After sometime, an equilibrium level of the temperature will be reached. From that time on, the temperature will stay there "forever." The dim temperature will be almost constant at 50°C. No changes will be felt as time passes by (Fig. (5.6)).

With this last example, we end our tour of sensing the workings of the Second Law on a system of dice. We will analyze the underlying principle of all of these experiments in the next chapter, and its relevance to the real world of the Second Law in Chapter 7.

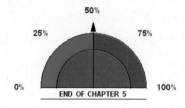

Finally, Grasp it with Your Common Sense

After having experienced various manifestations of the Second Law with dice, it is time to pause, analyze and rationalize what we have learned so far. We recall that we have *observed* different phenomena with different mechanisms. We shall see in the next chapter that some of these examples (color, taste and scent) have counterparts in real experimental systems. Other examples cannot be performed using particles (single particles do not emit sound waves; and the temperature that we feel with the tip of our fingers is a result of the distribution of velocities. One just cannot assign a temperature to each molecule). There are of course many more examples. The central question that concerns us in this chapter is: What are the features that are common to all the phenomena that we have observed in the experiments described in Chapters 4 and 5? The phenomena discussed in this chapter are essentially the same as those in Chapters 4 and 5, except that N is very large, much larger than the largest value of N discussed previously.

The three questions we need to ask ourselves are:

1) *What* is the common *thing* that we have observed changing towards something we called the "equilibrium line," and that once we have reached that line, no further changes can be observed?

2) *How* did we achieve that change? What is the essential aspect of the mechanism that leads us from the initial to the final state?

3) *Why* has that change occurred in only one direction, and no further changes are observed once we reach the equilibrium line?

We shall discuss these questions for the simple prototype experiment with a two-outcome dice. Our conclusions will apply to all the other types of dice that we have discussed in the previous chapters and it will likewise apply to real experiments as discussed in the next chapter.

We recall that our system consists of N dice. The outcome of tossing any die is either "0" or "1," with equal probability $1/2$. We have prescribed the rules by means of which, we change the configuration of the dice. We have seen that the rules can be changed. What is important is that there is at least one element of randomness in the rules; either we choose a die at random and change the outcome of the die deterministically, or we choose a predetermined order for selecting the die and throwing it to obtain a new outcome randomly, or we do both steps randomly.

We have defined a *specific* configuration or a specific event as a precise specification of the outcomes of each individual die. The exact configuration of the four dice in Fig. (6.1) is: The first die (red) on the left shows "0"; the second die (blue) next to the first shows "1"; the third one (green) shows "1"; and the last one (yellow) shows "0." This specification gives us a complete and detailed description of the system.

We have used the term dim configuration, dim state or dim event for a less detailed description of the system. In a dim event, we specify only the number of "ones" (or equivalently the number of "zeros") regardless of which specific die carries the number "one" or which carries the number "zero."

Fig. (6.1)

Thus, a dim description of the system in Fig. (6.1) is simply 2, or dim-2.

When we go from the specific configuration to the dim configuration, we disregard the *identity* of the die (whether it is red or blue, or whether it stands first or second in the row). We say that the dice, in this description, are indistinguishable. Here, we *voluntarily* give up the knowledge of the identity of the dice in passing from the specific to the dim description. In the real world, atoms and molecules *are* indistinguishable by nature, and that is an important difference between dice and atoms; this will be discussed in the next chapter.

Two characteristics of the dim configuration should be noted carefully.

First, for any dim description: "There are n ones in a system of N dice," the number of *specific* descriptions corresponding to this dim description grows with N. As a simple example, consider the dim description "There is a *single* ($n = 1$) one in a system of N dice." Here, there are exactly N *specific* configurations, comprising dim-1.

Second, fixing N, the number of specific configurations constituting the same dim configuration grows as n grows from zero up to $N/2$ (depending on whether N is even or odd, there are one or two maximal points, respectively). We have already seen that kind of dependence in Chapter 4. We give another example here for a system of fixed $N = 1000$, where n changes from zero to N (Fig. (6.2)).

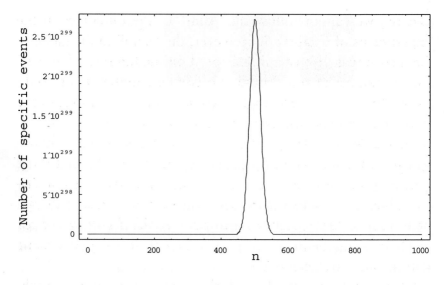

Fig. (6.2)

It is important to take note of these two trends. It requires some counting that might be tedious for large N and n, but there is no inherent difficulty and no sophisticated mathematics required, just plain and simple counting.

Once we have defined the specific configuration and the dim configuration, we can answer the first question we posed in the beginning of this section, viz., "*What* is the thing that changes with each step in the games in Chapters 4 and 5, which is also common to all the games?"

Clearly, the thing that you have *observed* changing is different in each experiment. In one, you see the color changing from yellow to green; in the others, you see changes in taste, smell, temperature, etc. All of these are different manifestations of the same underlying process. We are now interested in the *common* thing that changes in all of the experiments we have discussed in the last two chapters.

Let us go back to the "0" and "1" game discussed in Chapter 4. There, we have followed the sum of the outcomes.

Clearly, we cannot follow the "sum" of the outcomes in the experiments of Chapter 5. However, the "sum" in the game in Chapter 4 was also calculated based on the number of "ones" (or the number of zeros) in the experiment. Similarly, we could assign "one" and "zero" to the two colors, the two tastes, or the two temperatures, and track the number of "ones" in each of the games discussed in Chapter 5. This is fine, but not totally satisfactory. We need to find a proper *name* to describe that thing that is common to all the experiments, and give it a numerical value. This number increases until it reaches an equilibrium value. Let us tentatively call this number the *d-entropy* (*d* for dice) or simply dentropy. For the moment, this is just a name with no attached meaning yet.

This will be fine for this specific game. However, you might raise two objections to the description. First, we know that (real) entropy always increases. In this example, the dentropy will go upwards if we start from an "all-zeros" configuration. But what if we start with an "all-ones" configuration? The dentropy will dive downwards. This contradicts our knowledge of the behavior of the real entropy. The second objection you might raise is this: What if the dice have three different outcomes, say "zero," "one," and "two," or even non-numerical outcomes such as tones *A*, *B* and *C*, or three or four colors, or perhaps an infinite range of colors, velocities, etc.? What number should we be monitoring?

The two objections can be settled by first noting that the *thing* that we observe or feel in each specific game is one thing, but the *thing* that we monitor is another.

In our simple game of dice, we have been monitoring the number of "ones." This number increases steadily only if we start with an all-zeros configuration. Had we started with an all-ones configuration, the number of ones will *decrease* steadily towards the equilibrium line. We shall now see that with a simple transformation we can monitor the thing that always increases towards the equilibrium line.[1] To do this, we shall need the symbol of the absolute magnitude as defined in Chapter 2.

Instead of the number of "ones" denoted by n, we can monitor the number $|n - N/2|$. Recall that N is the number of dice, and n is the number of "ones." Hence, this quantity measures the deviation, or the "distance" between n and half of the number of dice. We take the absolute value so that the distance between, say $n = 4$ and $N/2 = 5$, is the same as the distance between $n = 6$ and $N/2 = 5$. What matters is how *far* we are from the quantity $N/2$,[2] which, as you recall, is the equilibrium line.

When we start with an all-zeros configuration, we have $n = 0$ and hence, $|n - N/2| = N/2$. When we start with all-ones, we have $n = N$ and $|n - N/2| = N/2$. In both cases, the distances from $N/2$ are the same. As n changes, this quantity will tend to change from $N/2$ to zero. So we have a quantity that almost **always** *decreases* from whatever initial configuration we start with.[3] Once we get to the minimal value of $|n - N/2| = 0$, we are at the equilibrium line. Check that for say, $N = 10$ and all the possible values of n. If you do not like to monitor a decreasing number, take the negative values of these numbers,[4] i.e., $-|n - N/2|$. This will increase steadily from $-N/2$ to zero. If you do not like to monitor negative numbers, take $N/2 - |n - N/2|$. This number will increase steadily from whatever initial configuration towards the maximum value of $N/2$ in all

---

[1]There is no real need to do that. However, we do that to make the behavior of the dentropy consistent with the behavior of the real entropy as discussed in Chapter 7.
[2]An equivalent quantity would be the square of $n - N/2$, i.e., $(n - N/2)^2$.
[3]Recall that we used the words "always" and "never" in the sense discussed at the end of Chapter 4.
[4]Again, it is not essential to do so. Recall the *H-theorem*. There, the *H-quantity* also decreases towards equilibrium (See Chapter 1).

the cases.[5] As you can see, by a simple transformation, we can define a new quantity that "**always**" *increases* with time (or with the number of steps). But this quantity does not answer the second objection. It is good for the game with only two possible outcomes. It cannot be applied to the more general case where the die has three or more outcomes. Thus, we need to search for a quantity that is common to all the possible experiments of the kind discussed in Chapters 4 and 5.

We shall now construct a new quantity that will always increase and which is valid even for the more general cases. There are many possibilities of choosing such a quantity. We shall choose the quantity that is closest to the quantity called entropy. To do this, we need the concept of information, or more precisely, the mathematical measure of information.

The quantity we shall choose to describe the *thing* that changes is the "missing information." We shall denote it by *MI* (for now, *MI* is an acronym for "missing information" but later in the next chapter, it will also be identified with the concept of entropy).

This quantity has several advantages in that it describes quantitatively, what the *thing* that changes in the process is. First, it conforms to the meaning of information as we use it in our daily lives. Second, it gives a *number* which describes that thing that changes from any initial state to the final state, and for any general game. It is always a positive number and increases in our dice games as well as in the real world. Finally and most importantly, it is the quantity which is common to all dice games and therefore is suitable to replace the tentative term dentropy.

---

[5]If you like, you can also "normalize" this quantity by dividing it by $N/2$ to obtain a quantity that starts at zero and end up at one.

It will also be identical to the quantity that is common to all *physical* systems, i.e., entropy.[6]

The qualitative definition is this: We are given a configuration described *dimly*, e.g., "there are n "ones" in the system of N dice." We are not informed of the *exact* configuration. Our task is to find out which the *specific* configuration given is.[7]

Clearly, from the knowledge of the dim configuration alone, we cannot infer the exact or the specific configuration; we need more *information*. This information is called the missing information or *MI*.[8] How do we get that information? By asking binary questions. We can define the *MI* as the number of binary questions we need to ask in order to acquire that information, i.e., knowledge of the specific configuration.

We have seen in Chapter 2 that the missing information is a quantity that is defined in such a way that it is independent of the way we acquire this information. In other words, it does not matter what strategy we use. The *MI* is "there" in the system. However, if we use the *smartest* strategy, we could identify the *MI* as the average *number* of binary questions that we need to ask. Thus, in order to use the *number* of binary questions to measure the amount of the *MI*, we must choose the smartest procedure to ask questions as described in Chapter 2. It is clear that the larger the *MI*, the larger will be the number of questions we need to ask. Let us look at the calculations in a few examples. Given the information that "there is a single 'one' in a system

---

[6]Since we are only interested in changes in the entropy, it is enough to determine the entropy up to an additive constant; also, a multiplicative constant that determines the units of our measure of entropy. See also Chapters 7 and 8.

[7]The definition also applies to the more general case where the dim description is "there are n_A showing face A, n_B showing face B, etc., in a system of N dice."

[8]Note that we can use either of the terms "information," or "missing information." The first applies to the information that is in the system. We use the second when we want to ask how much information we would need to acquire in order to know the specific state or the specific configuration.

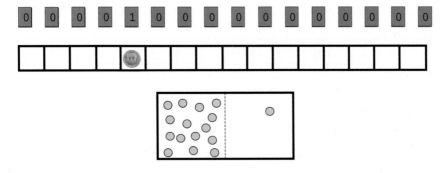

Fig. (6.3)

of 16 dice" (Fig. (6.3)) how many questions will we need to ask to obtain the *MI* in order to know the exact or the specific configuration?

This is exactly the same problem as that of finding the hidden coin in 16 equally probable boxes (Fig. (2.9)). So we shall proceed with the same strategy (see Chapter 2) and ask: Is it in the upper part? If the answer is yes, we choose the rhs and ask again: Is it in the rhs (of the four left boxes). If the answer is no, we simply choose the half which contains the coin. With this strategy, we will find the coin with exactly three questions. Clearly, if we have to find a single "one" in a larger N, say, $N = 100$ or 1000, the *MI* will be larger and we will need to ask more questions. Try to calculate the number of questions to be asked where the single "one" is in $N = 32$ and $N = 64$ dice.[9]

Next, suppose we are given "'two ones' in a system of 16 dice" (Fig. (6.4)). To acquire the *MI* in this case, we need to ask more questions. First, we can ask questions to locate the first "one," then we will ask the same type of questions to locate the second "one" in the remaining 15 dice.

---

[9]One can prove that this number is $\log_2 N$, i.e., the logarithm to the base 2 of the number of dice. Note that N is the number of dice. In this example, it is also the number of specific configurations.

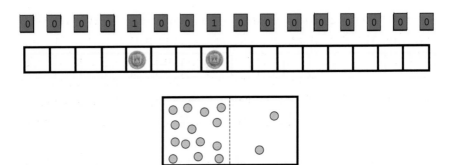

Fig. (6.4)

Clearly, for a fixed N, the number of questions required to be asked in order to obtain the required information increases with n. The larger n is, the more questions we will need to ask to locate all of the "ones" (or all of the hidden coins).

The number of questions required can be easily calculated for any n and N. We simply start by asking questions to determine where the first "one" is in the N dice, where the second "one" is in the remaining $N-1$ dice, where the third "one" is in the remaining $N-2$ dice, and so on, until we have all the information required on the n dice. This is true for any n provided it is smaller than $N/2$.

You should also be smart enough not only in choosing the best strategy but also in choosing which of the outcomes are to be located. If n is larger than $N/2$, it would be better for us to switch to asking for the locations of the "zeros" instead of the "ones." For instance, if we are given three "ones" in a system of four dice, the *MI* is exactly the same as if we were given a single "one" in a system of four dice. In this case, it will be smart to ask questions to locate the single "zero" in the first case, or the single "one" in the second case. By asking two questions, the entire, exact configuration can be determined.

Thus, the quantity *MI* increases with N for a fixed n. For a fixed N, the *MI* increases with increasing n, from $n = 0$ (i.e., with all "zeros," we need to ask zero questions) to $n = N/2$ (or to $(N+1)/2$ if N is odd) and then decreases when n increases beyond $n = N/2$. When we reach $n = N$, again *MI* is zero (i.e., when all are "ones," the number of questions we need to ask is zero too).[10]

A similar procedure can be prescribed for cases where there are more than two outcomes. This is slightly more complicated but it is not essential in order to understand the Second Law.

We now have a quantity referred to as *MI*, which is a *number* that describes the missing information we need in order to specify the exact configuration whenever we are given only the dim configuration. This number can be easily computed for any given n and N.[11]

Let us go back to our game in Chapter 4, where we monitored the *sum* of the outcomes or the number of *ones* in the evolution of the game. Instead of these two equivalent numbers, we shall now monitor the *MI* at each step. This is a more general quantity (it can be applied to any number and any type of outcomes), and it always increases (no matter from which state we start), reaching a maximum at some point (in this case $n = N/2$). Most importantly, this is the quantity which can be shown to be identical to the entropy of a real system.

One should realize that *MI* is a quantity we *choose* to monitor the evolution of the game. This is one of many other possible quantities we can choose (other possibilities being the

---

[10]Note that the *MI* discussed in this paragraph is strictly an increasing function of N, and of n (*for* $n < N/2$). In realizing an experiment with dice, the *MI* that we monitor behaves similarly to the curves shown in Chapter 4, for the sum as a function of the number of steps.

[11]The number of questions is $\log_2 W$, where $W = \frac{N!}{(N-n)!n!}$ is the total number of configurations to place n "ones" (or coins) in N dice (or N boxes). Note that this number is symmetric about $n = N/2$, at which point W is maximum.

number of "ones," the sum of the outcomes, or the quantity $N/2 - |n - N/2|$). The *thing* that changes is the dim *state* or the dim *configuration* of the system. The number we have assigned to these states is only an index that can be measured and monitored. The same index can be applied to any of the experiments we have carried out in Chapters 4 and 5, where the outcomes are not numbers, but colors, tones, scents, tastes or temperatures. All these are different *manifestations* of the same underlying process; the change from a dim configuration having a small index (e.g., *MI* or the sum) to a dim configuration having a larger index. The only thing that remains is to give this index a *name* – nothing more. For the moment, we use the name *MI* which means "missing information" and clearly does not elicit any mystery. Having found a name for the index we are monitoring, a name that also has the meaning of information,[12] we can discard the tentative term dentropy. We can use the term *MI* instead. We shall see later that *MI* is essentially the same as the entropy of the system.[13]

Let us now move on to the next question posed at the beginning of this chapter. *How* do we get from the initial to the final state?

The answer is very simple for these particular games of dice. We have prescribed the *rules* of the game. The simplest rules are: *choose a die at random, throw it to get a new random outcome and record the new dim configuration.* This answers the question "how?"

We have also seen that we have some freedom in choosing the rules. We can choose a die in some orderly fashion (say from left to right, or from the right to left, or any other prescribed way),

---

[12] This aspect has already been discussed by Shannon (1948). A more detailed exposition of this topic is given by Ben-Nain (2007).

[13] Actually, we are never interested in the absolute entropy of the system. What matters, and what is measurable, is only the difference in entropy.

and then throw the die to obtain a random new outcome. Or, we could choose a die at random and then change the outcome in a predetermined fashion; if it is "0," it will be changed to "1," and if it is "1," it will be changed to "0." There are many other rules that can be applied, for instance, choose two (or three, or four, etc.) dice at random and throw them. The evolution of the game will be slightly different in detail for each step, but the gross view of the evolution will be the same as we have witnessed in the experiments in Chapter 5. The important thing is that the rules should give each of the dice a "fair" chance to change, and to have a random element in the process. Within these limits, we have many possible rules to achieve the change.

It should be noted that we can easily envisage a non-random rule where the evolution will be very different; for instance, if we choose a die in an orderly fashion, say from left to right, then change the face of the die in a predetermined way from "0" to "1" or from "1" to "0." Starting with an *all-zeros* configuration, the system will evolve into an *all-ones* configuration, and then back to *all-zeros* and so forth, as shown in the sequence below.

$$\{0,0,0,0\} \rightarrow \{1,0,0,0\} \rightarrow \{1,1,0,0\} \rightarrow \{1,1,1,0\} \rightarrow$$
$$\{1,1,1,1\} \rightarrow \{0,1,1,1\} \rightarrow \{0,0,1,1\} \rightarrow \{0,0,0,1\} \rightarrow$$
$$\{0,0,0,0\} \rightarrow \{1,0,0,0\} \rightarrow \cdots$$

In such a case, the evolution of the system is every different as compared with the evolution shown in Chapters 4 and 5.

We could also prescribe rules that elicit no change at all (choose a die at random and do not change its face), or change the entire configuration from *all-zeros* to *all-ones* (choose a die in an orderly fashion and always change from "0" to "1," and from "1" to "1"). These rules are not of interest to us. As we shall discuss in the next chapter, these rules have no counterparts in the physical world (see Chapter 7).

We conclude that the answer to the "how" question is very simple. All we need is to *define* the rules in such a way that they contain some element of randomness, and to give each of the dice a "fair" chance to change from one outcome to another.

Finally, we address the last and the most important question. *Why* does the system evolve from a low *MI* to a high *MI* (or why did the "sum" or the number of "ones" increase steadily towards the equilibrium line in the games in Chapter 4)?

The answer to this question lies at the very core of the Second Law of Thermodynamics. Although the answer given here to this question is strictly pertinent to our simple game of dice, we shall see that it is also valid for the real physical processes.

As in any law of physics, there are two possible answers to the question "Why." One can simply answer with "that is the way it is," nothing more, nothing less. There is no way of understanding any of Newton's laws of motion in a *deeper* way. A moving ball when left uninterrupted by any force, will continue moving in a straight line and at constant velocity forever. Why? There is no answer to this question. That's the way it is. That is how nature works. There is no *logical reason*, nor explanation. In fact, this law sounds "unnatural" since it conflicts with what we normally observe in the real world. A non-scientific minded person who read the manuscript of this book was surprised to hear that such a law exists exclaiming "Everyone knows that a moving ball, left uninterrupted, will eventually stop." This law is not based on, and cannot be reduced to, common sense. In fact, most of the quantum mechanical laws are even counterintuitive and certainly do not sound to us as logical or natural (the reason is probably because we do not "*live*" in the microscopic world, and quantum mechanical effects are not part of our daily experiences).

The second answer is to seek for a deeper underlying principle or an explanation of the law. This is what people have attempted to do for decades for the Second Law. The Second Law of Thermodynamics is unique in that one *can* give an answer to the question "Why" based on logic and common sense (perhaps the only other law of nature that is also based on common sense is Darwin's law of natural evolution[14]).

We have seen in the previous chapters that there are many different manifestations of the (essentially) same process (and many more in the real world). Though in the various experiments, we monitored different manifestations, e.g., in one game, the number of "ones"; in another, the number of "yellows"; yet in another, the number of "sweet" dice — we have decided to use the same index, *MI* to follow the evolution of all these different games. These are different *descriptions* of essentially the same underlying process. "The system evolves towards more greenness," "the system evolves towards the larger sum," "the system evolves towards the larger *MI*," and so on.[15] All of these are correct descriptions of what goes on, but none can be used to answer the question "Why." There is no law of nature which states that a system should change towards more *greenness*. This is obvious. Neither is there a law of nature which states that a system should change towards more *disorder* or more *MI*.

If I have given you the answer to the "Why" question as, "because nature's way is to go from order to disorder, or from a low *MI* to a large *MI*," you can justifiably continue to ask.

---

[14]Here, "common sense" is strictly in the sense of logic. The theory of evolution, until very recently, was very far from being "common sense." It was only after the discovery of DNA, and the ensuing understanding of the mechanism of evolution on a molecular level, that the theory became common sense.

[15]A more common statement is: "The system evolves towards more disorder." We shall comment on this in Chapter 8.

Why? Why does the system change from a low to a high degree of disorder, or from a low to a high *MI*? Indeed, there is no such law. The things we have monitored are good for *describing* but not for *explaining* the cause of this evolution. To answer the question "Why," we need an answer that does not elicit, a new "Why" question.

The answer to the question "Why" (for all the processes we have observed so far, and indeed for all *real* processes as well) is very simple. In fact, it can be reduced to nothing but common sense.

We have seen that in each game, starting with any initial configuration, the system will proceed from a dim configuration consisting of a smaller number of specific configurations, to a new dim configuration consisting of a larger number of specific configurations. Why? Because each specific configuration is an *elementary event*, and as such, it has the same probability. Therefore, *dim* configurations that consist of a larger number of elementary events have a larger probability. When *N* is very large, the probability of the *dim events*, towards which the system evolves, becomes extremely high (nearly one!).[16] This is tantamount to saying that:

Events that are expected to occur more frequently, will occur more frequently. For very large N, more frequently equates with always!

This reduces the answer to the question "Why" to a mere tautology. Indeed, as we have seen in Chapter 2, probability is nothing but common sense, so is the answer to our question "Why."

---

[16]Note that I wrote "*dim-events*" not the "*dim-event*" corresponding to the equilibrium line. The latter has a maximal probability but it is not one! The dim events referred to here are the dim event corresponding to the equilibrium line together with its immediate vicinity. More on this in Chapter 7.

The changes we have observed in *all* the experiments are from dim events of lower probabilities to dim events of higher probability. There is nothing mysterious in this finding. It is simply a matter of common sense, nothing more.

It is also clear from this point of view, that the increase of *MI* (as well as of the entropy — see next chapter) is not associated with an increase in the amount of material or of energy.

You will be wondering, if entropy's behavior is nothing but common sense, why all these talks about the deep mystery of the Second Law? I shall try to answer this question in Chapter 8. For the moment, we are still in the world of dice. I suggest that you choose a number N, say 16 or 32, or whatever, and run the game either mentally or on your PC, according to one of the rules described in Chapters 4 and 5. Follow the evolution of the configurations and ask yourself *what* the thing that changes is, *how* it changes, and *why* it changes in the particular way. Your answers will be strictly relevant to this specific game of dice, but as we shall see in the next chapter, the answers are also relevant to the behavior of entropy in the real world.

Translating from the Dice-World to the Real World

In Chapter 6, I have assured you that if you understood the evolution of the dice games and if you could answer the questions "What," "How," and "Why," you are almost through in the understanding of the Second Law; all that is left to do is to show that what you have learned in the dice-world is relevant to the real world.

In this chapter, we shall translate the language of dice into the language of two real experiments. We shall start with the simplest, well-known and well-studied experiment: the expansion of an ideal gas.

To make the translation easier, let us redefine the dice game from Chapter 4 by considering dice having the letter R etched on three faces, and the letter L on the remaining three faces. Thus, instead of "0" and "1," or yellow and blue, or sweet and sour, we simply have two letters R and L. (R and L stand for "right" and "left", but at the moment it can be interpreted as any two outcomes of the dice, or of the coins.) We start with an all-Ls system and run the game according to the rules as prescribed in Chapter 4. We can monitor the number of "Rs," or the number of "Ls," or the missing information, MI. Thus, in this system, we shall find that after some time, both the number of "Rs" and the number of "Ls" will be almost equal to $N/2$, where N is the

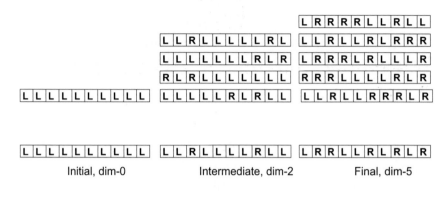

Fig. (7.1)

total number of dice. We show three stages of this game with $N = 10$ in Fig. (7.1).

Note that the initial state (dim-0) is unique. There is only one *specific* state that belongs to this dim state. In the intermediate state (dim-2), we have many possible *specific* states ($10 \times 9/2 = 45$) for the particular dim state which is described in the figure. The last state (dim-5) is again one specific state out of many possible specific states ($10 \times 9 \times 8 \times 7 \times 6/5! = 252$). This is the maximal dim state. In Fig. (7.1), we have indicated some of the specific configurations that comprise the dim states.

7.1. The Correspondence with the Expansion Process

Consider the experimental system depicted in Fig. (7.2). We have two compartments of equal volume separated by a partition. The compartment on the right is called R, and the compartment on the left is called L. We start with N atoms of, say argon, all the atoms being in the left compartment L. As long as we do not remove the partition, nothing noticeable will occur. In fact, on the microscopic level, the particles are incessantly jittering randomly, changing their locations and velocities. However, on the macroscopic level, there is no measurable change. We can

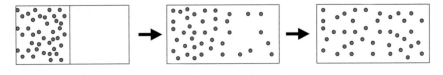

Fig. (7.2)

measure the pressure, temperature, density, color or any other properties, and we shall find a value that does not change with time or with location. We say that the system is initially contained in compartment L and is at an equilibrium state.[1] If we remove the partition, we set the Second Law in motion! We shall now *observe* changes. We can monitor the color, pressure, density, etc. We shall find that these quantities will change with time and location. The changes we *observe* are always in one direction; atoms will move from compartment L to compartment R. Suppose we monitor the density or color in L. We will observe that the density steadily decreases (or the intensity of the color, if it has color will diminish) with time. After some time, the system will reach an equilibrium state and you will no longer observe any further changes in any one of the parameters that you have been monitoring. This is referred to as the new equilibrium state. Once you have reached this state, the system will stay there "*forever.*" It will "*never*" return to the initial state. This process is one relatively simple manifestation of the Second Law.

In this specific process, we started from one equilibrium state (all atoms in L) and progressed to a new equilibrium state (the atoms are dispersed within the entire space of L and R). Let us repeat the experiment in a slightly different manner. This will ease the translation from the dice game to the real world.

---

[1]Here we refer to the *equilibrium state* in the thermodynamic sense. It should be distinguished from the equilibrium line. This distinction will be made clear later on in this chapter.

Suppose that instead of removing the partition as in Fig. (7.2), we only open a very small hole between the two compartments such that only one molecule, or very few can move from L to R at any given interval of time. If we open and close this little door at short intervals of time, we shall proceed from the initial state to the final state, exactly as in Fig. (7.2), but in this case, we shall proceed in a series of intermediate equilibrium states.[2]

We show three stages along this process in Fig. (7.3).

Let us now make the following correspondence between the world of dice and the real world of the expanding gas in the present experiment.

Each die corresponds to one specific atom, say, one argon particle. The faces of the dice in the last experiment of "R" and "L," correspond to a specific atom in the R and the L compartment, respectively.

We make the correspondence between the two "worlds" more specific for the case $N = 2$, shown in the two upper panels in Fig. (7.4). Note that in this correspondence, we distinguish between the particles (red and blue). On the lower panel

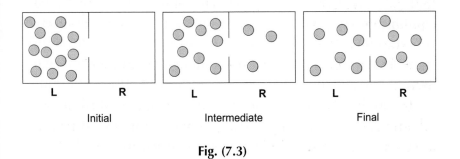

| L | R | L | R | L | R |

Initial   Intermediate   Final

Fig. (7.3)

---

[2]This kind of process is referred to as a quasi-static process. If the door is small enough, we do not need to open and close it every time an atom passes. The system is not in an equilibrium state, but the measurable quantities like densities, temperature, color, etc., will change very slowly, as if we were passing through a sequence of equilibrium states.

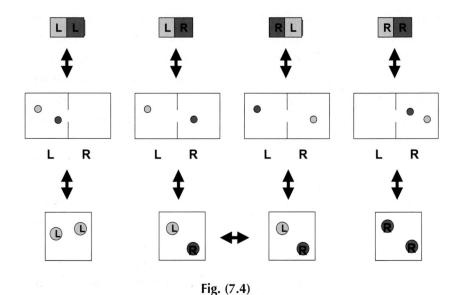

Fig. (7.4)

of Fig. (7.4), we have also added the correspondence with the assimilation process that I shall describe later in this chapter.

Next, we define a specific configuration as a complete specification of which particle is in which compartment. In contrast to the dice-world, where we can distinguish between the dice (although we disregarded this information when we monitored only the quantities that are pertinent to the dim configuration), here, the particles are indistinguishable from the outset. Therefore, we do not have to give up any information. The indistinguishability is a property of the atoms; it is something that nature imposes on the particles. The dice could be identical in all aspects, yet they are distinguishable in the sense that we can monitor each die. If we shake 10 identical dice, we can monitor one specific die, and at any point in time we can tell where this specific die comes from. In defining the dim configuration of dice, say five *R*s in 10 dice, we *can* distinguish between all the different *specific* configurations. We can tell *which* die carries an *R*, and *which* carries an *L*, as is clear from Fig. (7.1) or

Fig. (7.4). We cannot do this in an atomic system. All we can know or measure is the *number* of atoms in R and not which specific atoms are in R or in L. Thus in the system depicted in Fig. (7.4), we cannot distinguish between the two specific states: "blue particle in L and red particle in R" and "blue particle in R and red particle in L." These two states coalesce into one dim state: "one particle in L and one particle in R."

We refer to a configuration as dim when we specify only the *number* of particles in R (this will correspondingly specify the number of particles in L), disregarding the detailed specification of which atom is in L or in R. Clearly, each dim configuration consists of many specific configurations (except for the all-Rs or all-Ls configurations). It should be noted again that all we can measure or monitor is the *total* number of particles in R or any quantity that is proportional to this number (e.g., intensity of color, scent, density, pressure, etc.) We cannot "see" the specific configuration as we did in the dice-games.

Because of its crucial importance, we will once again describe the distinction between a dim configuration and the corresponding specific configurations (Fig. (7.5)).

In order to clarify the difference between the specific configurations, we have assigned different colors to the particles. In a system of particles, consisting of atoms or molecules, there are no labels (or colors) to distinguish between two identical particles. Hence, all we can see or measure is the dim configuration as shown on the left hand side of Fig. (7.5). It should be stressed, however, that although we cannot distinguish between the specific configurations, these do contribute to the *probabilities* of the dim configuration. Each *specific* configuration here is assumed to be equally probable.[3] Therefore, the probability of

---

[3] See, however, the discussion of fermion and boson particles in Chapter 2.

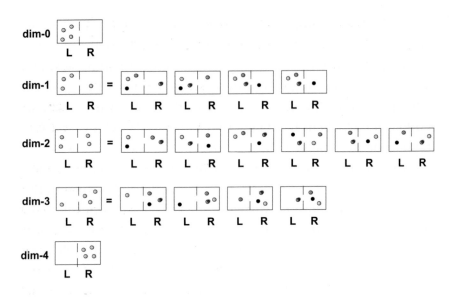

Fig. (7.5)

each dim configuration is the sum of the probabilities of the specific configurations that comprise the dim event. You can think of all the specific events on the right hand side of Fig. (7.5) as coalescing into one dim event on the left hand side. The probabilities of the five dim events are: $1/16, 4/16, 6/16, 4/16, 1/16$.

We are now in a position to answer the three questions of "What," "How" and "Why" for the real experiment. As in the case of the dice games, the question "What is the thing that changes?" has many answers. For instance, we can observe the intensity of the color as it evolves during the expansion. We can monitor the change in density, taste or scent with time. We can even calculate the change in the approximate number of particles in each compartment with time. Initially, we start with N atoms in compartment L. As we open the little door or remove the partition, the number of atoms, n, in L steadily decreases with time up to a point when it stops changing.

The answer to the "What" question is *exactly* the same as the answer we have given to the "what" question in the previous chapter, *viz.*, it is the dim configuration that is changing. While that is changing, it carries along with it some properties that we can see, hear, smell, taste or measure with macroscopic instruments. Thus, to the question: "What is the thing that changes?" in this experiment can be answered if we compare this experiment with the analysis given in the dice game. The only difference is the interpretation of the term "configuration," that has been defined here in terms of particles in two different compartments of R and L, while in the dice case, configuration was specified in terms of two outcomes, R and L of the die. Once we have made the correspondence between the game of dice and the expansion of the gas, we can use the same answer to the question, "What is the thing that is changing?" We shall come back to the question of what the *best* quantity is, that describes the thing which is common to all the processes later on.

Now we move on to the next question: "How did we get from the initial to the final state?" In the game of dice, we prescribed the rules of the game. So the answer to the question, in the context of the dice game, was straightforward; the changes occur according to the prescribed rules. The answer to the same question is different for the real experiment. Here, in principle, the equations of motion govern the evolution of the locations and the velocities of all the particles. However, in a system of a very large number of particles, one can apply the laws of probability.[4] We can loosely say that if we start from some exact

---

[4]In classical mechanics, if we know the exact locations and velocities of all the particles at a given time, then, in principle, we could have predicted the locations and velocities of all the particles at any other time. However, this vast information cannot be listed, let alone solving some 10^{23} equations of motion. The remarkable success of statistical mechanics is a monumental witness to the justification of applying statistical arguments to a system composed of very large numbers of particles.

description of all the locations and velocities of all particles, after a short time, the system will lose that information. Due to the random collisions and the roughness of the walls, the evolution of the system can be more effectively described by the laws of statistics than by the laws of mechanics.[5]

Thus, we can effectively apply probabilistic arguments similar to the ones applied in a dice game, i.e., there is an element of *randomness* that gives any one of the particles a "chance" to move from L to R or from R to L. Therefore, the answer to the question of "how" is not exactly, but *effectively*, the same as in the case of the dice game. We have also seen that the precise rules in the dice game were not very important; what was important is that each die had a "fair" chance to change and in a random fashion. This argument applies to the real experiment of expansion of a gas as well, i.e., each atom or a molecule must have a "fair" chance to move from L to R or from R to L.

If we remove the element of randomness, then the system will not evolve according to the Second Law. In Chapter 6, we prescribed rules, the application of which results in either *no* change at all, or a change from "all zeros" to "all ones," i.e., oscillating between these two extreme configurations. Likewise, we can think of a system of molecules that will not evolve according to the Second Law.

Consider the two following "thought experiments." Suppose that all the particles were initially moving upward in a concerted manner as shown in Fig. (7.6a). If the walls are perfect planes, with no irregularities, no roughness and exactly perpendicular to the direction of motion of the atoms, then what we will observe

---

[5]There is a deep and difficult problem in attempting to predict the probabilities from the dynamics of the particles. These efforts which started with Boltzmann have failed. One cannot, in principle, derive probabilities from the deterministic equations of motions. On the other hand, it is well-known that a system of very large number of particles moving randomly shows remarkable regularities and predictable behavior.

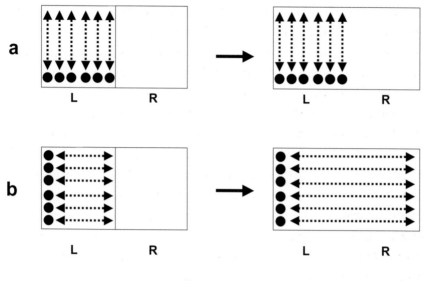

Fig. (7.6)

is that the particles will move upwards and downwards forever. Even after the partition is removed, all the particles that were initially in compartment L will stay in this compartment. The Second Law cannot "operate" on such a system.[6]

A second "thought experiment" is illustrated in Fig. (7.6b). Suppose again that we start with all the particles in L, but now the particles move in straight lines from left to right and from right to left. Initially, all the particles will move in concert and at the same velocity. The trajectory of all the particles will be the same, hitting the partition and bouncing back. If we remove the partition, the stream of particles will now move in concert from L to R and back from R to L, indefinitely. In both of these thought-experiments, there will be no evolution governed by the Second Law. In fact, such a process cannot be achieved in a real

---

[6]Such a system is of course impossible to realize. Realization of such an experiment is tantamount to knowing the exact location and velocity of each particle at any given time.

experiment. That is why we have referred to this process as a thought-experiment.

Clearly, any real walls must have some imperfections and even if we could have started with either of the above initially synchronised motions, very soon the laws of probability would take over and the Second Law would become operative.

It should be noted that in thermodynamics, we are not interested in the question of how a system moves from the initial to the final state. All that matters is the *difference* between the final and initial state. Here, we have looked through a magnifying glass on the details of the motion of individual particles to establish the correspondence between the rules of the dice game, and the rules of moving from L to R, or from R to L in the gas expansion. It is time to proceed to the next and most important question of "why!"

As we have noted in Chapter 6, the answer to the question "Why", given in the dice game, can also be applied to the case of the gas expansion. The gas will proceed from dim configurations of lower probability to dim configurations of high probability. Here, a specific configuration means a detailed specification of which particle is in which compartment. A dim configuration (which is the only one that we can monitor) is "how many particles are in compartment R." If you have understood the arguments presented in Chapters 4–6 regarding the evolution of the system from the initial state to the equilibrium state, then you would also understand the evolution of the physical system described in this chapter. We have seen that even for some 10^4 or 10^5 dice, the probability of returning to the initial state is negligibly small and concluded that once the system reaches the vicinity of the equilibrium line, it will stay there "**forever**," "**never**" returning to the initial state. The argument is *a fortiori* true for a system of 10^{23} dice or particles.

As in the case of the dice game, we have stressed that there is no law of nature that says that the system should evolve from yellow to green, or from order to disorder, or from less *MI* to more *MI*. All of these are either observable manifestations or means of monitoring the evolution of the system. The fundamental reason for the observed evolution of the system is obvious and self evident. Any system will always spend more time in states of higher probability than states of lower probability. When N is very large, say on the order of 10^{23}, "high probability" turns into "certainty." This is the essence of the Second Law of Thermodynamics. It is also a basic law of common sense, nothing more.

7.2. The Correspondence with the Deassimilation Process

In drawing the correspondence between the evolution of the dice game and the gas-expanding experiment, I have completed my mission: to guide you in understanding the workings of the Second Law of Thermodynamics. However, I would like to draw another correspondence between the dice game and a physical experiment. This additional case will not add anything new to your understanding of the Second Law, but will perhaps add one more example of a spontaneous process governed by the Second Law of Thermodynamics. My motivation for presenting it is mainly for aesthetic reasons. Let me explain why.

Any spontaneous process involving an increase in entropy is driven by the same law of common sense, i.e., events that are more probable will occur more frequently. We have seen only one such physical process, a spontaneous expansion of gas. There are, of course, more complicated processes, such as a chemical reaction, mixing of two liquids, splattering of a falling egg, and many more. It is not always easy to define precisely the states of the system on which the Second Law is operative. In teaching thermodynamics, it is customary and quite instructive to classify processes according to the type of states which are involved in the processes. In terms of information, or rather in terms of the missing information, we sub-classify processes according to the type of information that is lost. In the expansion process, each particle was initially located in a smaller volume V. After the expansion, it became located in a larger volume $2V$, hence, it is more "difficult" to locate the particle or equivalently, we have less information on the *locations* of the particles. In a process of heat transfer between two bodies at different temperatures, there is a more subtle change in the amount of information. Before the process, the particles in the hot body are characterized by one distribution of energies (velocities) and the particles in the cold body by another distribution of velocities. After contact and equilibration, there is a single distribution of the velocities of all the particles in the two bodies. We shall discuss this process again below. In more complicated processes such as the splattering of an egg, it is difficult to define the types of information that are lost; among others, it may be information on the location, velocity, orientation etc. That is a very complicated process, sometimes beyond our ability to describe.

In the dice game, we had N identical dice; each could be in two (or more) states, say, "0" and "1," or yellow and blue, or "R" and "L." In the expansion process, we have made the correspondence between the two outcomes of the dice with the two *locations* of the particles of the same kind (say, argon atoms). This is fine. We can always denote an atom in compartment R as an R-*atom* and similarly, an atom in compartment L as L-*atom*. This is formally correct but aesthetically unsatisfactory since the inherent identity of the atoms did not change in this process. In other words, we have made the correspondence between the *identity* of the outcome of the dice and the *location* of the particle.

Let me present a new experiment which is again driven by the Second Law, but where the correspondence between the dice game and the physical process is more plausible and more aesthetically satisfying. We shall make the correspondence between the *identity* of the outcome of the dice and the *identity* of the particles.

Going through this process also affords us a little "bonus": we can imagine real experiments where we can follow the color, scent, or taste of the system as it evolves with time.

Consider a molecule having two isomers, say, *Cis* and *Trans*, of a molecule, schematically depicted in Fig. (7.7).

Starting with pure *Cis*, the system might not change for a long time. If we add a catalyst (the analogue of removing the partition), we would observe the spontaneous change from the pure *Cis* form into some mixture of *Cis* and *Trans* forms. Statistical mechanics provides a procedure for calculating the ratio of the concentrations of the two isomers at equilibrium. In this particular reaction, we can identify two different kinds of causes for the entropy changes, equivalently, two kinds of informational changes; one is associated with the *identity* of the molecule, while the other is associated with the re-distribution

Fig. (7.7)

of the energy over the internal degrees of freedom of the two species.

There is one particular example of a chemical reaction in which only the *identity* of the molecules changes (the internal degrees of freedom are the same for the two species). This is the case of two enantiomers, or two optically active molecules. These are two isomers that have exactly the same structure and the same chemical constituent; the only difference is that one is a mirror image of the other. Here is an example of such a molecule (Fig. (7.8)).

Let us call these isomers d and l (d for dextro and l for levo).[7]

These two molecules have the same mass, same moment of inertia, same internal degree of freedom, and the same set of energy levels. Thus, when we perform a "reaction" involving the transformation from d to l form, the only change in the system is in the number of indistinguishable particles. Let us perform the following experiment.[8] Start with N molecules of the d form.

Alanine *l*   Alanine *d*

Fig. (7.8)

---

[7]The two isomers rotate the plane of a polarized light to the right (dextro) or to the left (levo).

[8]This process has been called deassimilation, i.e., the reversal of the assimilation process defined as a process of the loss of identity of the particles. In some textbooks, this process is erroneously referred to as mixing. For more details, see Ben-Naim (1978, 2006).

Place a catalyst that induces a spontaneous transformation from d to l, or from l to d. It can be proven that at equilibrium, we will find about $N/2$ of the *d-form* and about $N/2$ of the *l-form*.[9] One can also calculate the entropy change in this process and find that it is exactly the same as in the expansion process that we have discussed before. However, the kind of "driving force" is different, and in my opinion, the correspondence between the dice game and this process is more satisfactory and more "natural."

To see this, we note that in both of the experiments, we have made the correspondence:

a specific die ↔ a specific particle

In the expansion experiment, we have also made the correspondence between:

a specific *identity* of the outcome of a die

↕

a specific *location* of the particle

In the second experiment, referred to as deassimilation, the correspondence is between:

a specific *identity* of the outcome of a die

↕

a specific *identity* of the particle

Thus, whereas in the expansion process the evolution from the initial to the final state involves changes in the *locational*

---

[9]This is intuitively clear. Since the two isomers are identical except for being a mirror image of each other, there is no reason that at equilibrium there will be one form in excess of the other. Exactly for the same reasons that in the expansion process, there will be nearly the same number of particles in R and in L (presuming that the volumes of the two compartments are the same).

information about the particles, in the deassimilation process, on the other hand, there is a change in the *identity* of the particles.[10] This is the same *type* of informational loss we have monitored in the dice game.

The correspondence between the dice and the particles for this process is shown on the lower panel of Fig. (7.4).

In both the dice game and the deassimilation processes, there is an evolution involving a change of the *identity* of the particles. We start with N, all-zeros dice in one case, and with all *d-form* in the real experiment. After some time, $N/2$ of the dice have changed into "ones" in one case, and $N/2$ of the particles acquire a new *identity*, the l form, in the other case.[11] This correspondence is smoother and more natural than the one between the dice game and the expansion process. The evolution of the system can be exactly described in the same way as we have done for the expansion of the gas. Merely replace R and L by d and l, and you will understand the evolution of this experiment. As previously shown in Figs. (7.1) and (7.3), we also show in Fig. (7.9) the three stages of the process of deassimilation and the correspondence with both the expansion experiment and dice game.

The answers to the "What" and "Why" questions are exactly the same as for these two processes. The answer to the "how" question is slightly different.[12] However, as we have noted

---

[10]It should be noted that in an ideal gas system, there are only two kinds of information; locational and velocities. The identities of the particles do not consist of a new kind of information. However, change of identity does contribute to the change in information (for more details, see Ben-Naim 2007).

[11]Any time we refer to $N/2$ particles, we mean in the vicinity of $N/2$.

[12]Instead of the probability of any atom hitting the door in the partition and crossing from R to L or from L to R, there is a probability of any isomer of acquiring enough energy (by collisions), to cross from the *d-form* to the *l-form* and vice versa, or alternatively to hit the catalyst, which will induce the transformation from one isomer to the other.

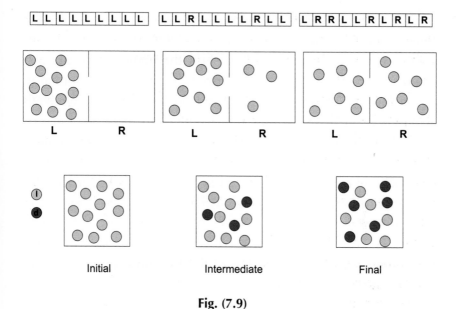

Initial   Intermediate   Final

Fig. (7.9)

above, the question of "how" is not important to an understanding of the Second Law. What is important is only the difference in entropy between the initial and final states.

Now for the bonus: in Chapter 5, we discussed some hypothetical processes where the color, taste or scent was monitored in a system of dice.

These processes can, in principle, be made real. Suppose we have two isomers having different structures so that they have different colors, scent or tastes. Performing the experiment of the spontaneous isomerization, we could follow the color change from blue to green (as in the first example in Chapter 5); or from scent *A* to a mixture of scents *A* and *B* (as in the second example); and from a sour taste to a sweet and sour taste (as in the third example). These changes can be monitored continuously in a homogenous system.[13]

---

[13] We could also achieve the same effect by mixing two different gases, as in Fig. (1.4). In this case, the color (or the scent or taste) will be monitored continuously but not homogeneously throughout the system, as in the process described in this section.

It is difficult to find a physical experiment analogous to the tone-change (fourth example) since single particles do not emit sound waves. It is impossible to find the analogue of the last example of Chapter 5. Temperature is a complex phenomenon depending on a continuous distribution of velocities. We shall discuss a process involving temperature changes in Section 7.5.

It should be clear by now what the thing is that changes (which we call entropy) and why it changes in the particular way, whether it is in the dice game, the expansion process or the deassimilation process (Fig. (7.9)).

7.3. Summary of the Evolution of the System towards the Equilibrium State

Consider again the simple experiment as described in Fig. (7.2). We start with $N = 10^{23}$ particles in L. A *specific* configuration is a detailed list of which particle is in L and which is in R. A *dim* configuration is a description of how many particles are in L and in R. As long as the partition is in place, there is *no change* in the dim configuration nor in the specific configuration.[14] The system will not evolve into occupying more states if these are not accessible.

Now, remove the barrier. The new total number of specific states is now 2^N; each particle can either be in L or R. The total number of states, $W(total)$, is fixed during the entire period of time that the system evolves towards equilibrium.

---

[14]There is a subtle point to note here. The information we are talking about is where the particles are; in L or in R. We could have decided to divide the system into many smaller cells, say four cells in R and four cells in L. In such a case, the *information* of the initial state is different since we shall be specifying which particle is in which *cell*; so is the information of the final state. However, the *difference* between the information (as well as the entropy) is independent of the internal divisions, provided we do the same division into cells in the two compartments. [For further details, see Ben-Naim, (2006, 2007).]

Clearly, once the barrier is removed, changes will occur and we can observe them. It could be a change in color, taste, smell or density. These are different manifestations of the same underlying process. What is the thing that changes and is common to all the manifestations?

The thing that changes is the *dim* state or *dim* configuration or *dim* event, and there are various ways of indexing, or assigning to these dim states a *number* that we can plot so as to be able to follow its numerical change. Why does it change? Not because there is a law of nature that says that systems must change from order to disorder, or from a smaller *MI* to a bigger *MI*. It is not the *probability* of the dim states that changes (these are all fixed).[15] *It is the dim state itself that changes from dim states having lower probabilities to dim states having higher probabilities.*

Let us follow the system immediately after the removal of the barrier, and suppose for simplicity, that we open a very small door that allows only one particle to pass through in some short span of time. At the time we open the door, the dim state consists of only *one* specific state belonging to the post-removal condition (i.e., when there are zero particles in *R*). Clearly, when things start moving at random (either in the dice game or in the real gas colliding with the walls and once in a while hitting the hole), the first particle that passes through the hole is from *L* to *R*, resulting in the new state, dim-1. As we have seen in great detail in Chapter 4, there is a high probability of the system either moving up or staying at that level, and there is a very small probability of the system going downwards to a lower dim state. The reason is quite simple. The probability of any

---

[15]There is an awkward statement in Brillouin's book (1962) that "the probability tends to grow." That is probably a slip-of-the-tongue. What Brillouin probably meant is that *events* of low probability are evolving into events of high probability. The probabilities themselves do not change.

particle (any of the N) crossing the border between L and R is the same. Let us call that probability p_1 (which is determined by the speed of motion, the size of the door, etc.). Whatever p_1 is, the probability of moving from dim-1 to dim-0 is the probability of a single particle in R crossing over to L, which is p_1. On the other hand, the probability of crossing from dim-1 to dim-2 is $(N-1)$ times larger than p_1 simply because there are $(N-1)$ particles in L, and each has the same chance of crossing from L to R. The same argument can be applied to rationalize why the system will have a higher probability of proceeding from dim-1 to dim-2, from dim-2 to dim-3, etc. Each higher dim state has a larger number of specific states and therefore a larger probability. As we have seen in Chapters 3 and 4, this tendency of a system going upwards is the strongest initially, becoming weaker and weaker as we proceed towards dim-$N/2$, which is the equilibrium line. This equilibrium line is the dim state with highest probability since it includes the largest number of specific states.

One should be careful to distinguish between the number of specific states belonging to dim-$N/2$ and the total number of states of the system which is $W(total)$. These are two different numbers. The latter is the *total* of all the specific states included in all the possible dim states. For instance, for $N = 10$, we have[16]

$$
\begin{aligned}
W(Total) &= W(\text{dim-0}) + W(\text{dim-1}) \\
&\quad + W(\text{dim-2}) + \cdots \\
&= 1 + 10 + 45 + 120 + 210 + 252 + 210 + 120 \\
&\quad + 45 + 10 + 1 = 2^{10}
\end{aligned}
$$

---

[16]In general, this equality can be obtained from the identity $(1+1)^N = 2^N = \sum_{n=0}^{N} \frac{N!}{n!(N-n)!}$.

As we have discussed in Chapters 2 and 3, the probability of the dim event is just the sum of the probabilities of the specific events that comprise the dim event.

For example, for dim-1 of the case $N = 10$, we have 10 specific events. The probabilities of each of these 10 specific events are equal to $(1/2)^{10}$. The probability of the *dim*-event is simply 10 times that number, i.e.,

$$\text{Prob(dim-1)} = 10 \times (1/2)^{10}$$

For dim-2, we have $10 \times 9/2 = 45$ specific events. All of these have the same probability of $(1/2)^{10}$. Therefore, the probability of dim-2 is:

$$\text{Prob(dim-2)} = 45 \times (1/2)^{10}$$

In the table below, we have listed the probabilities of all the dim events for the case $N = 10$. Note again the maximum value at dim-$N/2$ or dim-5.

Dim-Event	Number of Specific Events	Probability
dim-0	1	$1/2^{10}$
dim-1	10	$10/2^{10}$
dim-2	45	$45/2^{10}$
dim-3	120	$120/2^{10}$
dim-4	210	$210/2^{10}$
dim-5	252	$252/2^{10}$
dim-6	210	$210/2^{10}$
dim-7	120	$120/2^{10}$
dim-8	45	$45/2^{10}$
dim-9	10	$10/2^{10}$
dim-10	1	$1/2^{10}$

Figure (7.10) shows the number of specific events for each dim event, for different N ($N = 10$, $N = 100$ and $N = 1000$).

In the lower panel, we show the same data but in terms of probabilities of the dim events.

It is clear from Fig. (7.10) that as N increases, the *number* of specific states belonging to the maximal dim states also becomes very large. However, the *probability* of the maximal dim state *decreases* with N. Thus, as N increases, the probability distribution of the dim events is spread over a larger range of values. The apparent sharpness of the distribution as shown in the lower panel of Fig. (7.10) means that deviations from the maximal dim state, in *absolute* values, are larger, the larger N is. However, deviations from the maximal dim states, *relative* to N, become smaller, the larger N is.

For instance, the probability of finding deviations of say $\pm 1\%$ from the maximal dim state becomes very small when N is very large.

We next calculate the probability of finding the system in any of the dim states between, say, $N/2 - N/100$ to $N/2 + N/100$, i.e., the probability of the system being found *around* the maximal dim state, allowing deviations of $\pm 1\%$ of the number N. This probability is almost one for $N = 10,000$ (see Fig. (7.11)). With

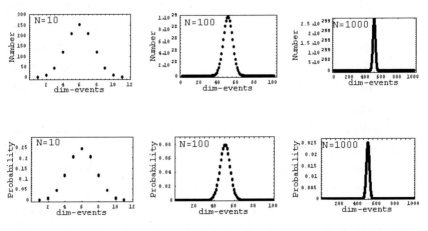

Fig. (7.10)

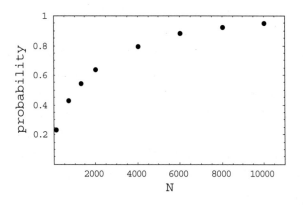

Fig. (7.11)

N on the order of 10^{23}, we can allow deviations of 0.1% or 0.001% and still obtain probabilities of about one, of finding the system at or near the equilibrium line.

What we have found is very important and crucial to understanding the Second Law of Thermodynamics. The probability of the maximal dim-$N/2$ *decreases* with N. However, for very large N, on the order of 10^{23}, the system is almost certain (i.e., probability nearly one) to be in one of the dim states within very narrow distances from the equilibrium dim state. When $N = 10^{23}$, we can allow extremely small deviations from the equilibrium line, yet the system will spend almost all the time in the dim states within this very narrow range about the equilibrium line.

We stress again that at the final equilibrium state, the *total* number of specific states is 2^N, and all of these have *equal* probabilities and therefore each of these *will* be visited with an extremely low frequency $(1/2)^N$. However, the dim states (altogether there are $N + 1$ dim states) have *different* probabilities since they include different numbers of specific states. The dim states at and near the maximal dim-$N/2$ will have a probability of nearly one, i.e., nearly the same probability of being in any one of the total specific states.[17]

It is now time to clarify the relationship between the equilibrium *line*, and the experimental equilibrium *state* of the system. At equilibrium, an experimental system spends the *larger* fraction of time at the equilibrium *line*, larger, compared with all the other dim states. This does not imply *all* the time. The experimental or thermodynamic equilibrium *state* of the system is the state for which *all* the W(*total*) specific events are accessible and have equal probability. However, since we cannot distinguish between dim states that are very near the equilibrium line, the system will spend nearly *all* the time in the vicinity of the equilibrium *line*. Deviations will occur. There are two types of deviations. When the deviations are very small, they do occur and very frequently, but we cannot *see* them. On the other hand, we *can* see large deviations, but they are so rare that we "*never*" see them. Therefore, the equilibrium *state* of the system is (almost) the same as the dim event of the equilibrium line and its immediate vicinity.

Because of the importance of this point, we will repeat the argument in slightly different words.

Consider the following experiment. We start with two compartments having the same volume V; one contains N *labeled* particles $1, 2, \ldots, N$; the second contains N, labeled particles $N + 1, N + 2, \ldots, 2N$.[18] We now remove the partition between the two compartments. After some time we ask: "What is the

---

[17]By "near" the maximal dim-$N/2$, we mean near in the sense of a very small percentage of N, say 0.001% of N. These states are so close that experimentally they cannot be distinguished. If we start with a system having exactly $N/2$ in each compartment separated by a partition, then remove the partition, the number of states increases from $\frac{N!}{(N/2)!(N/2)!}$ to 2^N. This is a huge change in the number of states. However, for large $N \approx 10^{23}$, we will not be able to see any change in the system. Each compartment will contain nearly, but not exactly, $N/2$ particles.

[18]Assuming for the moment that particles can be labeled.

probability of observing *exactly* the initial state?" The answer is: 2^{-2N}. The second question is: "What is the probability of observing *exactly* N particles in each compartment regardless of the labels?" This is a far larger probability.[19] However, this probability *decreases* with N, as can be seen from Fig. (7.10). The probability of observing the dim-N state, although very large compared with any of the other single dim states, is still very small. The actual experimental observation is not the *exact* dim-N, but a group of dim states in the neighborhood of the equilibrium line dim-N. This neighborhood consists of all the dim states between which we cannot distinguish experimentally.[20]

What about other more complicated processes?

We have discussed in detail the expansion of a gas where we chose only one parameter to describe the evolution of the events, i.e., the particle being either in L or R. In the deassimilation experiment, we also chose only one parameter to describe the events, i.e., the particle being either an *l* or a *d* form. All we have said about the expansion process can be translated almost *literally* to the deassimilation experiment. Simply replace "being in L" or "being in R," by "being an *l-form*" or "being a *d-form*."

There are, of course, more complicated processes involving many more "parameters" to describe the events; a molecule can be in this or that place, can have one or another velocity, can be one or the other isomer (or conformer in larger molecules), etc.

In order to *understand* the Second Law, it is sufficient to understand one process — the simplest, the best. That is what we have done. The principle is the same for all the processes;

---

[19]This probability is $(2N)!/(N!)^2 \times 2^{-2N}$.

[20]We cannot distinguish between very close dim states, say between dim-N and dim-N + 1, or dim-N + 1000. This indistinguishability is different from the indistinguishability between specific-states, belonging to the same dim state. In the former, the indistinguishability is in *practice*; in the latter, the indistinguishability is in *principle*.

they differ only in the details. Some are easy and some, more difficult, to describe. Some processes are so complicated that we still do not know how to describe them. Sometimes, we even do not know how many parameters are involved in the process. Let us describe briefly some processes of increasing degrees of complexity.

7.4. Mixing of Three Components

Suppose we start with three different gases, N_A of *A-molecules* in a volume V_A, N_B of *B-molecules* in a volume V_B and N_C, *C-molecules* in a volume V_C. (Fig. (7.12)).

We remove the partitions between the three compartments and watch what happens. If the molecules have the same color, we will not be able to see anything but we can measure the densities or the concentrations of each type of molecules at each point and record the changes. If the molecules have different colors, tastes or scents, we can follow the change of color, taste or scent as they evolve after the partitions are removed.

But how do we describe with a single number the thing, the change of which we are observing? Even with these relatively simple experiments, the construction of the numerical "index" (which we need in order to record the evolution of the system) is not simple. First, we have to define the specific events of our system. A specific event might read like: "molecule 1 of type *A* in volume V_A, molecule 2 of type *A* in volume V_B,... molecule

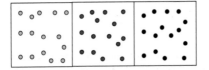 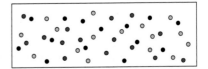

Fig. (7.12)

1 of type *B* in V_C, molecule 2 of type *B* in volume...etc." A very lengthy description indeed!

We recognize that there are many of these events among which we cannot distinguish, e.g., the specific event "molecule 1 of type *A* in V_A, and molecule 2 of type *A* in volume V_B, etc." is indistinguishable from the specific event "molecule 1 of type *A* in V_B and molecule 2 of type *A* in V_A, etc." The "etc." is presumed to be the same description of the rest of the locations of all the other molecules.

Next, we define the dim events, e.g., "one *A*-molecule in V_A, 15 *B*-molecules in V_B and 20 *C*-molecules in V_C, etc." Here, we disregard the labels of the molecules in each compartment. What matters is that one, *any* one, *A*-molecule is in V_A, and *any* 15, *B*-molecules in V_B, and *any* 20, *C*-molecules in V_C.

Having done that, we need to calculate the probabilities of all these dim events. This is not so easy in the general case. We do this by using the same assumptions we had in the expansion experiment, *viz.* the probabilities of all the specific events are equal and they are equal to $1/W(total)$. Therefore, for each dim event, we can calculate its probability simply by summing over all the probabilities of the specific events that comprise the dim event. However, in this case the dim event is not expressed by a single number, as was the case in the expansion process. In order to monitor the evolution of the dim events, we need a single number. That number is the *MI*, i.e., the number of binary questions we need to ask in order to find out at which of the specific states the system is, given only the dim state. This number, up to a constant that determines the units, is the entropy of the system which is defined for each dim state.[21]

---

[21] Note again that in order to define the entropy of each dim state, we have to open and close the little doors between the compartments at each point we want to calculate the entropy. In this manner we allow the system to proceed through a sequence of equilibrium states.

With this number, we can monitor the evolution of the system from the time we removed the partitions, until it reaches the final equilibrium state. If we do that, we should find that the *MI* has increased in this process.

7.5. Heat Transfer from a Hot to a Cold Gas

In Chapter 5, we described an "experiment" with dice involving temperature changes. We commented there that the experiment is extremely unrealistic. Here, we shall discuss a real experiment involving change of temperatures. This experiment is important for several reasons. First, it is one of the classical processes in which entropy increases. In fact, it was one of the simplest processes for which the Second Law was formulated (see Chapter 1). Second, it is important to demonstrate that the thing that changes in this process is the same as those in the other processes, i.e., the *MI*. Finally, to explain why we could not have devised a simple dice-game analog for this process.

Consider the following system. Initially, we have two isolated compartments, each having the same volume, the same number of particles, say argon, but at two different temperatures $T_1 = 50K$ and $T_2 = 400K$ (Fig. (7.13)). We bring them into contact (either by placing a heat-conducting partition between them, or simply removing the partition and letting the two gases mix). Experimentally, we will observe that the temperature of

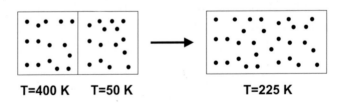

T=400 K   T=50 K   T=225 K

Fig. (7.13)

the hot gas decreases, while the temperature of the cold gas increases. At equilibrium, we shall have a uniform temperature of $T = 225K$ throughout the system.

Clearly, heat, or thermal energy is transferred from the hot to the cold gas. To understand how the entropy changes in this process, we need some mathematics. Here, I shall try to give you a qualitative feeling for the kind of entropy change involved in the process.

First, we note that temperature is associated with the distribution of molecular velocities. In Fig. (7.14), we illustrate the distribution of velocities for the two gases in the initial state. You will see that the distribution is narrower for the lower temperature gas, while it is broader for the higher temperature gas. At thermal equilibrium, the distribution is somewhat intermediate between the two extremes, and is shown as a dashed curve in Fig. (7.14).

What we observe experimentally is interpreted on a molecular level as the change in the distribution of molecular velocities. Some of the kinetic energies of the hotter gas is transferred to the colder gas so that a new, intermediate distribution is attained at the final equilibrium.

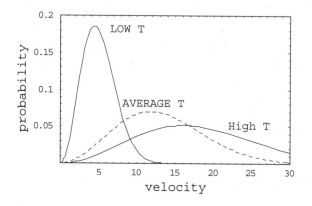

Fig. (7.14)

Now look at the two curves in Fig. (7.15), where we plotted the velocity distribution of the entire system before and after the thermal contact. Can you tell which distribution is more ordered or *disordered*? Can you tell in which distribution the *spread* of kinetic energy is more even, or over a larger range of values?[22] To me, the final distribution (the dashed curve) looks more ordered, and the distribution looks less spread-out over the range of velocities. Clearly, this is a highly subjective view. For this and for some other reasons discussed in the next chapter, I believe that neither "disorder," nor "spread of energy" are adequate descriptions of entropy. On the other hand, information or *MI* is adequate. Unfortunately, we need some mathematics to show that. I will only cite here a result that was proven by Shannon in 1948.[23] The final distribution of velocities is the one with the minimum *information* or maximum *MI*. Although this result cannot be *seen* from the curve, it can be proven mathematically.

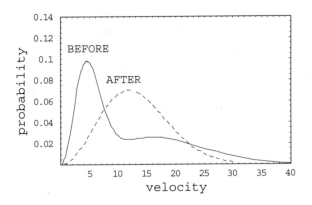

Fig. (7.15)

---

[22] Sometimes entropy is described as a measure of the spread of the energy. It should be noted that "spread" like "order" is sometimes appropriate but not always, as demonstrated in this example.

[23] Shannon (1948), section 20. More detailed discussion of this aspect of the MI may be found in Ben-Naim (2007).

Here again we get a single *number* that is related to the entropy, the evolution of which can be monitored.[24]

Finally, we recall that in the unrealistic game of dice that we have described in Chapter 5, we used dice having only two states, hot and cold, with equal probabilities. In order to make that experiment more realistic, we should have invented dice with an *infinite* number of faces (each corresponding to a different molecular velocity). Also, we have to change the rules of the game to describe the evolution towards equilibrium (we cannot change the velocity of each particle at random; the total kinetic energy must be conserved). All of these are difficult to implement in the dice game. Therefore, you should be cautious not to infer from that particular example of Chapter 5 anything that is even close to a real physical system.

In this example, we have followed experimentally one parameter, the temperature. However, the temperature is determined by an infinite number of parameters: all possible velocities. Of course, it is impossible to follow the velocity of each particle. What we do follow is the temperature which is a measure of the *average* velocities of the molecules. However, the *thing* that changes, the thing that we call entropy, is nothing but the change in the amount of missing information; a quantity that can be expressed in terms of the distribution of velocities, before and after the contact between the two gases.

Also the "driving-force" for this process is the same as the one for the expansion process, as well as for the dice-game; the system will proceed from a state of low probability to a state of higher probability.

---

[24]Provided we do the process quasi-statically, i.e., by letting the heat transfer proceed very slowly so that the equilibrium within each gas is nearly maintained at each stage. We can think of a small hole through which the gas can pass from one compartment to the other, similar to the case described in the expansion process. Alternatively, we can think of a very narrow, heat-conducting material, connecting the two systems.

Once we have more parameters to describe the events, the counting becomes more and more difficult. Very quickly we arrive at such processes (as the processes that take place from the time an egg hits the floor, until it reaches equilibrium, assuming that the egg and the floor are in a box and the box is isolated from the rest of the world) which are impossible to describe. In such processes, molecules change their locations, their distribution of velocities, and their internal states, such as vibrations, rotations, etc. It is virtually impossible to describe this, let alone calculate *the* probabilities of the various events. However, we believe that the principles that govern the changes that occur are the same as in the simple expansion experiment. In principle, we believe that there is a quantity that we call entropy which is better described as the missing information (MI) that changes in one direction for any process carried out in an isolated system. In other words, the Second Law is at work in conducting the multitude of events that unfold in all these processes.

It is tempting to include life processes in the same category of processes that are governed by the Second Law. However, I believe that at this stage of our understanding of "life", it would be premature to do so. It is impossible to describe and enumerate all the "parameters" that change in any life process. Personally, I believe that life processes, along with any other inanimate processes, are also governed by the Second Law. I shall make one further comment regarding life processes in the next chapter.

As we have seen, there are several "levels" in which we can describe what the thing is that changes. On the most fundamental level, the thing that changes is the specific *state*, or the specific *configuration* of the system; a specific die changes its face from 2 to 4. In the expansion process, a specific particle changes its location from say, "L" to "R", and in the deassimilation process, a specific particle changes, say, from "l" to "d." What we can monitor is something that is a property of the dim state. Each of these dim states contain many, many specific states, between which we either *do not care* to distinguish (like when we follow only the *sum* of the outcomes of N dice) or we *cannot* distinguish *in principle* (like which specific particle crossed from L to R). The thing that we monitor is something we can either measure (temperature, frequency of electromagnetic wave, density, etc.), or we can perceive with our senses (color, smell, cold or hot, etc). If we want a number or an index to monitor the evolution of the system, the best and the most general one is the missing information, or equivalently, the entropy.

Why does a system change from one dim state to another? Simply because the new dim state consists of many more specific states, and therefore, has a larger probability. Hence, the system spends a larger fraction of time in the new dim state.

And finally, why is it that when a system reaches the equilibrium state, it stays there "**forever?**" Simply because, the number of specific states which constitute the dim states near the equilibrium line is extremely large and each of these contributes the same probability.

Thus, a system always proceeds from a dim state of low probability to a dim state of higher probability. This is tantamount to saying that events which have high frequency of occurrence *will* occur with high frequency. This statement is nothing but common sense. When the number of particles is very large, the number of elementary events which comprise the same dim event is so large that the frequency of occurrence of the dim states near the equilibrium *line* which we referred to as the equilibrium *state* is practically one. Therefore, once this state is reached, it will stay there "**forever.**" This is exactly the same conclusion we have reached in the game of dice described in Chapter 6.

At this point, you have gained a full understanding of the reason for the evolution of two processes: the expansion of an ideal

gas and the deassimilation processes. If you wish, you can formulate your own version of the Second Law: *an ideal gas occu-*
pying an initial volume V will always expand spontaneously
to occupy a larger volume, say 2V. If the system is isolated,
*you will **never** see the reversed process occurring.* It is easy to
show that this formulation is equivalent to either Clausius' or
Kelvin's formulation of the Second Law (Chapter 1). To prove
that, suppose that your formulation of the Second Law will not
be obeyed, i.e., sometimes the gas occupying a volume $2V$ will
spontaneously condense into a volume V. If this occurs, you can
construct a simple contraption to lift a weight. Simply place a
weight on the compressed gas and wait until the gas expands
again. You can also use the spontaneous expansion to transfer
heat from a cold to a hot body. The trick is the same as the one
used to prove the equivalency of the Kelvin and the Clausius
formulation of the Second Law.

7.6. Test Your Understanding of the Second Law

Now that you understand the Second Law in dice games and you
have seen the translation from the dice language to the language
of real particles in a box, it is time that you test yourself with the
same set-up of an experiment as described in Fig. (7.2). Suppose
that you have never heard about the Second Law, but you know
and accept the following assumptions:[25]

1) Matter consists of a very large number of atoms or molecules,
   on the order of 10^{23}.

---

[25]Note that all this knowledge is supposed to be provided by physics. It is true,
though, that some of this knowledge was acquired after the formulation and the
study of the Second Law. Here, however, we presume that this information is *given*
to us in advance.

2) A system of 10^{23} atoms in an ideal gas consists of many specific states (or specific events or specific configurations) and these are assumed to have equal probabilities.

3) All the specific events can be grouped into dim events (like the ones in Fig. (7.5)).

4) Each dim event (except for dim-0 and dim-N) consists of a huge number of specific events among which we cannot distinguish (as those on the right hand side of Fig. (7.5)).

5) The probability of each dim event is the sum of the probabilities of *all* the specific events that constitute that dim event. The relative time spent by the system in each dim state is proportional to its probability.

6) There is one dim event that consists of a maximum number of specific events. Hence, the system spends a *larger* fraction of time at this maximal event.

7) We cannot distinguish between dim events that differ only by a small number of particles, say between dim-10^{23} and dim-$10^{23} \pm 1000$ or between dim-10^{23} and dim-$10^{23} \pm 10^6$.

The last assumption (7), is essential. In my opinion, this assumption (in fact, it is a fact!), is not emphasized enough in textbooks which explain the Second Law. Without this assumption, one can follow all the arguments in building up the Second Law but at the end, he or she might not reach the conclusion that entropy should stay *strictly* constant at equilibrium, and that entropy should not *strictly* change in one direction only (upwards). The fact that we do not *observe* any decrease in entropy at equilibrium, and do *observe* strictly increasing entropy in a spontaneous process is due to two reasons:

1. Small fluctuations do occur and very frequently, but they are *unobservable* and *un-measurable* because they are too small to be observed or to be measured.

2. Large fluctuations could be *observed* and *measured* but they are extremely rare and therefore ***never*** *observable* and *measurable*.

Now I will ask you questions and you will answer. Check your answers against mine. These are given in brackets.

We start with a system of N particles in a volume $V(N$ is very large, on the order of 10^{23}. A simpler system for illustration is shown in Fig. (7.2). Initially, the partition dividing the two compartments does not allow the particles to cross.

Q: What will you see?

A: (Nothing — any measurable quantity will have the same value at each point in the system, and this quantity does not change with time.)

Next, we open a small door between the left (L) and right (R) compartments. It is small enough so that only one particle can hit-and-cross the door within a short interval of time, say $t = 10^{-8}$ seconds.[26] At this interval of time there exists a probability, denoted p_1, that a *specific* particle will hit-and-cross the door. Since the atoms are identical, the same probability p_1 applies to *any* specific particle.

Q: What is the probability of *any* of the N particles hitting-and-crossing the door within the time interval t?

A: (Clearly, since we have assumed that the door is small enough so that no two particles can cross within the time interval t, the probability of particle 1 crossing is p_1, the probability of particle 2 crossing is p_1 ... and the probability of the N-th particle N crossing is p_1. Since all these events are disjoint, the probability of *any* particle crossing is N times p_1, or $N \times p_1$.)

---

[26]This is not essential, but it is easier to think of such an extreme case when only one particle at a time, can cross the door either way.

Q: Right. What will happen in the first interval of time *t*?

A: (Either a particle will cross from *L* to *R*, or nothing will happen.)

Q: Correct. Suppose we wait another time *t*, and another time *t* until the first particle crosses. Which particle will cross over?

A: (I do not know *which* particle will cross over, but I am sure that whichever particle crosses over, it must be from *L* to *R*.)

Q: Why?

A: (Simply because there are no particles in *R* so the first to cross over must come from *L*.)

Q: Correct. Now wait for some more intervals of time until the next particle crosses over. What will happen?

A: (Since there are now $N - 1$ particles in *L* and only one particle in *R*, the probability of a particle, *any* one of the $N - 1$ from *L* to hit-and-cross is much larger than the probability of the single particle crossing from *R* to *L*. So it is far more likely [with odds of $N - 1$ to 1] that we shall observe the second particle also crossing from *L* to *R*.)

Q: Correct. Let us wait again for the next particle to cross over. What will happen?

A;   (Again, since the relative probabilities of a particle crossing from *L* to *R* and from *R* to *L* is $N - 2$ to 2, it is much more likely that a particle will cross from *L* to *R*.)

Q: Indeed, and what about the next step?

A: (The same, the odds are now $N - 3$ to 3 of a particle crossing from *L* to *R*, this is slightly less than in the previous step, but since $N = 10^{23}$, the odds are still overwhelmingly in favor of a particle crossing from *L* to *R*.)

Q: What will be the next few millions or billions of steps?

A;   (Again, the answers are the same for each time. If there are
n in R and $N - n$ in L, and if n is very small compared with $N/2$,
millions or billions or trillions is still very small compared with
10^{23}, there will be a higher probability of a particle crossing
from L to R than from R to L.)

Q: What happens when n becomes equal or nearly equal to
$N/2$?

A: (The odds to cross from L to R are about $(N - n){:}n$, which
for n about $N/2$ means $\frac{N}{2}{:}\frac{N}{2}$, or equivalently, the odds are now
nearly 1:1.)

Q: So, what will happen next?

A: (What will *happen* is one thing and what I will see is
another. What will *happen* is that on the average, there will
be the same number of particles crossing from L to R as from R
to L. What I will *see* is nothing changing. If n deviates from $N/2$
by a few thousands, or a few millions of particles, I will not be
able to detect such a small deviation. If, however, a very large
deviation occurs, then I could possibly *see* that, or *measure* that,
but that kind of deviation is very unlikely to happen, so it will
never be observed.)

Q: So what will you see or measure from now on?

A: (No changes will be observed or measured; the system will
reach its equilibrium state where the number of particles in L
and R are about equal.)

Q: You have passed the test. One more question to check if
you have understood the deassimilation process. Suppose we
start with N molecules, all in the *d-form*. We introduce a tiny
piece of a catalyst. Whenever a molecule, *any* molecule, hits the
catalyst, there is a probability p_1 that it will change either from
d to l or from l to d. What will you observe in this system?

A: (Exactly the same answers as given before; just replace *L* by the *d-form*, and *R* by the *l-form*. Instead of a little door that allows crossing from *L* to *R*, or from *R* to *L*, the catalyst allows the change from *d* to *l*, or from *l* to *d*. With this translation from the *L* and *R* language to the *d* and *l* language, all the answers to your questions will be identical.)

Well, I think you have now understood two examples of how entropy behaves in a spontaneous process. You have passed the test, and I have accomplished my mission. If you are interested in reading some of my personal speculations, you are welcome to read the next chapter.

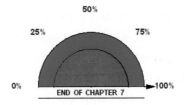

Reflections on the Status of the Second Law of Thermodynamics as a Law of Physics

If you have followed me so far and have reached this last chapter, you must feel comfortable with the concept of entropy and with the Second Law. If you throw a pair of (real) dice many times, and find that the *sum* = 7 appears on the average more than any other sums, you should not be surprised. If you throw one hundred simplified dice (with "0" and "1"), you should not be puzzled to find out that the sum of the outcomes will almost always be about 50. If you throw a million simplified dice, you should not be mystified to find out that you will "**never**" get the *sum* = 0 or the *sum* = 1,000,000. You know that both of these results are *possible* outcomes, but they are so rare that you can play all your life and will not witness even once that particular result. You will not be mystified because you have thought about that and your common sense tells you that events with high probability will be observed more frequently, while events with extremely low probability will "**never**" occur.

If you have never heard of the atomic constituency of matter and you watch a colored gas initially contained in one compartment of a vessel flowing and filling up the two compartments of the vessel, as shown in Fig. (8.1a); or two compartments with two different gases, say yellow and blue, transformed into a blend of homogenous green, as shown in Fig. (8.1b); or a hot

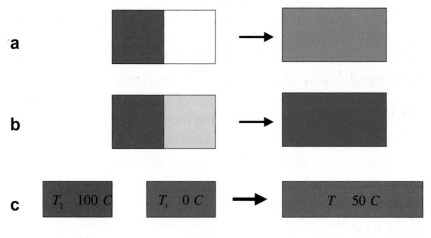

Fig. (8.1)

body at a temperature say $T_2 = 100°C$, when brought into contact with a cold body, say at $T_1 = 0°C$, cooled to a temperature somewhere in between T_1 and T_2, as shown in Fig. (8.1c), you *should* be mystified. Why did the blue gas flow from one chamber to fill the other chamber? Why were the two colored gases transformed into a single color? Why did the temperatures of the two bodies change into a single temperature? What are the hidden forces that propelled all these phenomena, and always in these directions and never in the opposite directions? Indeed, for as long as the atomic theory of matter was not discovered and accepted,[1] all of these phenomena were shrouded in mystery.

Mystery might not be the right word. Perhaps "puzzlement" will describe the situation better. The only reason for you to be puzzled is that you do not have any understanding of why these phenomena happen in the particular direction. But that is the same for any law of physics. Once you accept the law as a fact,

---

[1]By "discovered and accepted," I mean "*not yet* discovered and accepted." If matter *did not* consist of atoms and molecules, then there would have been no mystery none of the phenomena would have occurred. The Second Law as formulated within classical thermodynamics would not have existed at all.

you will feel that it is natural, and that it makes sense.[2] The same is true for the Second Law; the fact that these processes are so common in daily life, means that they are slowly and gradually being perceived as "natural" and "make sense."

If, however, you know that a gas consists of some 10^{23} atoms or molecules, jittering and colliding incessantly millions of times a second, then you know that the laws of probability will prevail, and that there is no mystery. There is no mystery in all these processes as much as there is no mystery in failing to win the "one million" prize in the last lottery.

I would like to believe that even if you encountered the words "entropy" and the "Second Law" for the first time in this book, you would be puzzled as to why the word "mystery" was associated with these terms at all. You will have no more reasons to cringe upon hearing the word "entropy," or to be puzzled by that unseen "force" that pushes the gas from one side to the other. There is also no need for you to continue reading this book. My mission of explaining the "mysteries of the Second Law" has ended on the last pages of Chapter 7, where you have reached a full understanding of the Second Law.

In this chapter, I take the liberty to express some personal reflections on the Second Law. Some of my views are not necessarily universally agreed upon. Nevertheless, I have ventured into expressing these views and taking the risk of eliciting the criticism of scientists whose views might be different and perhaps more correct than mine.

In this chapter, I shall raise some questions and shall try to answer them. I will begin with a relatively innocent question: "Why has the Second Law been shrouded in mystery for so long?" Is it because it contains a seed of conflict between

---

[2]Here, "makes sense" is used in the sense of being a common and familiar experience, not in the logical sense.

the time-reversal symmetry of the equations of motion, and the observed irreversibility of natural processes? Then I shall discuss a few other questions, the answers to which are still controversial. Is entropy really a measure of "disorder," and what does order or disorder of a system mean? How has "information" invaded a "territory" that used to harbor only physically measurable entities? Is the Second Law intimately associated with the arrow of time? What is the "status" of the Second Law of Thermodynamics *vis a vis* other laws of nature? Is it also possible that one day science will do away with the Second Law of Thermodynamics as it will be deemed a redundancy, a relic of the pre-atomistic view of matter that does not further enrich our knowledge of how nature works?

8.1. What is the Source of the Mystery?

In my opinion, there are several reasons which gave rise to the mystery enveloping the Second Law. The first, and perhaps the simplest reason for the mystery is the very word "entropy." Everyone is familiar with concepts like force, work, energy and the like. When you learn physics, you encounter the same words, although sometimes they have quite different meanings than the ones you are used to in everyday life. The amount of "work" that I have expended in writing this book is not measured in the same units of work (or energy) that are used in physics. Likewise, the "force" exerted on a politician to push for a specific law or a bill is not the same as the force used in physics. Nevertheless, the precise concepts of work and force as defined in physics retain some of the qualitative flavor of the meaning of these words as used in daily life. Therefore, it is not difficult to accommodate the new and more precise meaning conferred on familiar concepts such as force, energy or work. When you encounter, for the first time, a new word such as "entropy," it conjures up an air of mystery; it has a strange and uneasy effect on you. If you are not a student of physics or chemistry, and by chance hear scientists talking about "entropy," you will certainly feel that this concept is beyond you and *a fortiori* so, when you hear the scientists themselves referring to "entropy" as a mystery.

Leon Cooper (1968), right after quoting Clausius' explanation of his reasons for the choice of the word "entropy," comments[3]

> *"By doing this, rather than extracting a name from the body of the current language (say: **lost heat**), he succeeded in coining a word that meant the same thing to everybody: **nothing**."*

I generally agree with Cooper's comment but I have two reservations about it. First, the word "entropy" is unfortunately a misleading word. This is clearly different than meaning "nothing." Open any dictionary and you will find: "Entropy — Ancient Greek change, literary turn." Clearly, the concept of entropy is not "transformation," nor "change," nor "turn." As we have seen, entropy as defined in either the non-atomistic or the atomistic formulation of the Second Law is something that changes. But it is not the "transformation" that is transformed, nor the "change" that is changing, and certainly not the "turn" that is evolving.

My second reservation concerns the casual suggestion made by Cooper that "lost heat" could have been more appropriate. Of course, "lost heat" is a more meaningful term than "entropy." It is also in accordance with the universal meaning

---

[3]See Chapter 1, page 7. We again quote from Clausius' writing on the choice of the word "entropy." Clausius says: *"I propose, accordingly, to call S the entropy of a body after the Greek word "**transformation**."*

assigned to entropy as a "measure of the unavailable energy.[4] I will revert to this meaning assigned to entropy in Section 8.3 below.

Besides the unfamiliarity with a new concept that creates an air of mystery, there is a second reason for the mystery. The very fact that many authors writing on entropy *say* that entropy *is* a mystery, *makes* entropy a mystery. This is true for writers of popular science as well as writers of serious textbooks on thermodynamics.

Take for example a very recent book, brilliantly written for the layman by Brian Greene. He writes[5]:

> *"And among the features of common experience that have resisted complete explanation is one that taps into the deepest unresolved mysteries in modern physics, the mystery that the great British physicist, Sir Arthur Eddington called the arrow of time."*

On the next pages of the book, Greene explains the behavior of entropy using the pages of Tolstoy's epic novel *War and Peace*. There are many more ways that the pages of the said novel can fall out of order, but only one (or two) ways to put them in order.

It seems to me that the above quoted sentence contributes to perpetuating the mystery that is no longer there. In a few more sentences, Greene could have easily explained "entropy," as he explained so many other concepts of modern physics. Yet to me, it is odd that he writes: "...the deepest unresolved mysteries in modern physics," when I believe he should instead have written: "Today, the mystery associated with the Second Law no longer exists." There are many authors who wrote on the Second Law

---

[4]Merriam Webster's Collegiate Dictionary (2004).
[5]Greene (2004).

with the intention of *explaining* it, but in fact ended up propagating the mystery.[6]

Here is a classical example. Atkins' book on *The Second Law* starts with the following words[7]:

> *"No other part of science has contributed as much to the liberation of the human spirit as the Second Law of Thermodynamics. Yet, at the same time, few other parts of science are held to be recondite. Mention of the Second Law raises visions of lumbering steam engines, intricate mathematics, and infinitely incomprehensible entropy."*

What should one make of these opening sentences? I definitely do not agree with all the three quoted sentences. The first sentence is ambiguous. I failed to understand what the Second Law has got to do with "liberating the human spirit." However, my point here is not to argue with Atkins views on the Second Law. I quote these opening sentences from Atkins' book to demonstrate how each contributes to propagating the mystery. The first sentence elicits great expectations from the Second Law and presumably encourages you to read the book. However, these expectations are largely frustrated as you go on reading the book. The next two sentences are explicitly discouraging — "an infinitely incomprehensible entropy" does not whet your appetite to even try to taste this dish. In many textbooks on thermodynamics, the authors spend a lot of time discussing different manifestations of the Second Law, but very little on what is *common* to all these manifestations. Instead of selecting one or two simple examples of processes that are manifestations of the Second Law, the authors present a very large number of

---

[6] An exception is Gamov's book *One, Two, Three Infinity* that opens a section with the title *The Mysterious Entropy* but ends it with: *"and as you see, there is nothing in it to frighten you."*

[7] Atkins (1984).

examples, some of which are too complicated to comprehend. Reading all these, you cannot see the forest for the trees.[8]

In Chapter 7, we have discussed two relatively simple examples that demonstrate the workings of the Second Law. In each of these examples only one parameter changes. In the first, the change we observed was in the *locational* information, i.e., particles that are initially confined to a smaller volume, disperse and fill a larger volume. In the second example, the *identities* of the particles were changed. In the experiment on heat transfer from a hot to a cold body, it is the distribution of velocities that was changed. There are, of course, more complicated processes that involve changes in many parameters (or degrees of freedom). Sometimes, it is difficult to enumerate all of them. For instance, the processes that occur following the splattering of an egg involve changes of location, identities of molecules, distribution of velocities, orientations and internal rotations within the molecules. All of these complicate the description of the process, but the principle of the Second Law is the same. To understand the *principle*, it is enough to focus on one simple process, and the simpler, the better and the easier to understand.

Atkins' book devotes a whole chapter to "see how the Second Law accounts for the emergence of the intricately ordered forms characteristic of life."[9] In my opinion, this promise is not delivered. I have read Atkins' entire book, cover-to-cover, and I failed to "see how the Second Law accounts for the emergence of the intricately ordered forms characteristic of life."

These kinds of promises contribute to the frustration of the readers and discourage them from getting to grips with the Second Law.

---

[8]It is interesting to note that "entropy" and "the Second Law" feature in the titles of scores of books (see some titles of books in the bibliography). To the best of my knowledge, no other single law of physics has enjoyed that treat.
[9]Atkins (1984).

Life phenomena involve extremely complicated processes. Everyone "knows," scientists as well as non-scientists, that *life* is a complex phenomenon, many aspects of which, involving the mind and consciousness, are still not well understood. Therefore, discussing *life* in a book which is supposed to *explain* the Second Law leaves the reader with the impression that entropy, like life, is hopelessly difficult to understand and very mysterious.

It is true that many scientists believe that all aspects of life, including consciousness, are ultimately under the control of the laws of physics and chemistry, and that there is no such separate entity as the mind which does not succumb to the laws of physics. I personally believe that this is true. However, this contention is still far from being proven and understood. It might be the case that some aspects of life will require extension of the presently known laws of physics and chemistry, as was cogently argued by Penrose.[10] Therefore, in my opinion, it is premature to discuss life as just another example, fascinating as it may be, within the context of explaining the Second Law.

There are more serious reasons for the mystery that has befogged entropy. For over a century, the Second Law was formulated in thermodynamic terms and even after the molecular theory of matter has been established, the Second Law is still being taught in thermodynamics, employing macroscopic terms. This approach inevitably leads down a blind alley. Indeed, as my first lecturer correctly proclaimed (see Preface), there is no hope of understanding the Second Law *within* thermodynamics. To reach the light, you must go through the tunnels of statistical thermodynamics, i.e., the formulation of the Second Law in terms of a huge number of indistinguishable particles. If you go through the various different formulations of the Second Law within classical thermodynamics, you can prove the equivalence

---

[10]Penrose (1989, 1994).

of one formulation to some other formulations; you can show that the entropy that drives one process, say the expansion of a gas, is the same entropy that drives another process, say the mixing of two different gases. It is somewhat more difficult to show that it is also the same entropy that drives a chemical reaction, or mixing of two liquids. It is impossible to prove that it is the same entropy that causes the mess created by the splattering of an egg (yet we do assume that it is the same entropy and that one day, when the tools of statistical thermodynamics shall have been more powerful, we will be able to prove it). However, no matter how many examples you work out and prove that they are driven by the inexorably and the ever-increasing entropy, you will reach a blind alley. You can never understand what the underlying source of this one-way ascent of the entropy is. Thermodynamics does not reveal to you the underlying molecular events.

Had the atomic energy of matter not been discovered and accepted,[11] we would have never been able to explain the Second Law; it would have forever remained a mystery.

That was the situation at the end of the nineteenth century and at the beginning of the twentieth century. Although the kinetic theory of heat had succeeded in explaining the pressure, temperature, and eventually also the entropy in terms of the motions of atoms and molecules, these theories were considered to be *hypotheses*. Important and influential scientists such as Ostwald and Mach thought that the concept of the atom, and the theories based on its existence, should not be part of physics. Indeed, they had a point. As long as no one had "seen" the atoms directly or indirectly, their incorporation in any theory of matter was considered speculative.

---

[11]See footnote 1, page 187.

The situation changed dramatically at the beginning of the twentieth century. It was Einstein who contributed decisively to defeating the aether, and also paved the way for the atomists' victory. The acceptance of Boltzmann's molecular interpretation of entropy became inevitable (see Chapter 1).

But how come the mystery still did not vanish with the embracing of Boltzmann's interpretation of entropy? True, the door was then widely open to a full understanding of the ways of entropy and yet the mystery persisted.

I am not sure I know the full answer to this question. But I do know why, in my *own* experience, the mystery has remained floating in the air for a long time. The reason, I believe, involves the unsettled controversy which arose from the association of entropy with "disorder," with "missing information" and with the "arrow of time." I shall discuss each of these separately.

8.2. The Association of Entropy with "Disorder"

The association of entropy with disorder is perhaps the oldest of the three, and has its roots in Boltzmann's interpretation of entropy. Order and disorder are vague and highly subjective concepts, and although it is true that in many cases, increase in entropy can be correlated with increase in disorder, the statement that "nature's way is to go from order to disorder" is the same as saying that "nature's way is to go from low to high entropy." It does not explain why disorder increases in a spontaneous process. There is no law of nature that states that systems tend to evolve from order to disorder.

In fact, it is not true that, in general, a system evolves from order to disorder. My objection to the association of entropy with disorder is mainly that order and disorder are not well-defined, and are very fuzzy concepts. They are very subjective, sometimes ambiguous, and at times totally misleading. Consider the following examples:

In Fig. (8.2) we have three systems. On the left hand side, we have N atoms of gas in volume V. In the second, some of the N atoms occupy a larger volume $2V$. In the third, the N atoms are spread evenly in the entire volume $2V$. Take a look. Can you tell which of the three systems is the more ordered one? Well, one can argue that the system on the left, where the N atoms are gathering in one half of the volume, is more ordered than the system on the right, where N atoms are spread in the entire volume. That is plausible when we associate entropy with missing information (see below), but regarding order, I personally do not see either of the systems in the figures to be more ordered, or more disordered, than the other.

Consider next the two systems depicted in Fig. (8.3):

In the left system, we have N blue particles in one box of volume V and N red particles in another box of the same volume V. In the right, we have all the atoms mixed up in the *same* volume V. Now, which is more ordered? In my view, the left side is more ordered — all the blues and all the reds are

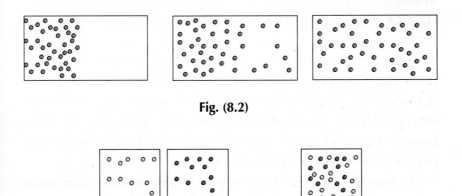

Fig. (8.2)

Fig. (8.3)

separated in two different boxes. On the right-hand side, they are mixed up in one box. "Mixed-up" is certainly a disordered state, colloquially speaking. In fact, even Gibbs himself used the word "mix-upness" to describe entropy. Yet, one can prove that the two systems mentioned above have *equal* entropy. The association of mixing with increase in disorder, and hence increase in entropy, is therefore only an illusion. The trouble with the concept of order and disorder is that they are not well-defined quantities — "order" as much as "structure" and "beauty" are in the eyes of the beholder!

I am not aware of any precise *definition* of order and disorder that can be used to validate the interpretation of entropy in terms of the extent of disorder. There is one exception, however. Callen (1985), in his book on thermodynamics, writes (p. 380):

> *"In fact, the conceptual framework of "information theory" erected by Claude Shannon, in the late 1940s, provides a basis for interpretation of the entropy in terms of Shannon's measure of **disorder**."*

And further, on the next page, Callen concludes:

> *"For closed system the entropy corresponds to Shannon's quantitative measure of the maximum possible disorder in the distribution of the system over its permissible microstates."*

I have taught thermodynamics for many years and used Callen's book as a textbook. It is an excellent textbook. However, with all due respect to Callen and to his book, I must say that Callen misleads the reader with these statements. I have carefully read Shannon's article "The Mathematical Theory of Communication," word-for-word and cover-to-cover, and found out that Shannon neither defined nor referred to "disorder." In my opinion, Callen is fudging with the definition of disorder in the quoted statement and in the rest of that chapter.

What for? To *"legitimize"* the usage of *disorder* in interpreting entropy. That clearly is not in accord with Shannon's writings. What Callen refers to as Shannon's definition of *disorder* is in fact Shannon's definition of *information*. In my opinion, Callen's re-definition of information in terms of disorder does not help to achieve the goal of explaining entropy. As we have seen in Chapters 2 and 6, the concept of information originated from a qualitative and highly subjective concept, has been transformed into a quantitative and objective measure in the hands of Shannon. As we have also seen, the distilled concept of "information" also retains the meaning of information as we use it in everyday life. That is not so for disorder. Of course, one can define disorder as Callen has, precisely by using Shannon's definition of *information*. Unfortunately, this definition of "disorder" does not have, in general, the *meaning* of disorder as we use the word in our daily lives, and has been demonstrated in the examples above.[12]

To conclude this section, I would say that increase in disorder (or any of the equivalent words) can sometimes, but not always, be associated with increase in entropy. On the other hand, "information" can *always* be associated with entropy, and therefore it is superior to disorder.

8.3. The Association of Entropy with Missing Information

Ever since Shannon put forward his definition of the concept of information, it has been found to be very useful in interpreting entropy.[13] In my opinion, the concept of missing information has not only contributed to our understanding of what is

---

[12]Furthermore, Shannon has built up the measure of information, or uncertainty, by requiring that this measure fulfill a few conditions. These conditions are plausible for *information*, but not for disorder. For further reading on this aspect of entropy see Ben-Naim (2007).

[13]See Tribus (1961) and Jaynes (1983) both dealing with the informational theoretical interpretation of entropy.

the *thing* that changes (which is called entropy), but it has also brought us closer to the last and final step of understanding entropy's behavior as nothing but common sense. This view, however, is not universal.

On this matter, Callen (1983, page 384) writes:

> *"There is a school of thermodynamics who view thermodynamics as a subjective science of prediction."*

In a paragraph preceding the discussion of entropy as disorder, Callen writes:

> *"The concept of probability has two distinct interpretations in common usage. 'Objective probability' refers to a **frequency**, or a **fractional occurrence**; the assertion that 'the probability of newborn infants being male is slightly less than one half' is a statement about census data. 'Subjective probability' is a measure of **expectation based on less than optimum information**. The (subjective) probability of a **particular yet unborn** child being male, **as assessed by a physician**, depends upon that physician's knowledge of the parents' family histories, upon accumulating data on maternal hormone levels, upon the increasing clarity of ultrasound images, and finally upon an educated, but still subjective, guess."*

As I have explained in Chapter 2 (in the section on "Conditional probabilities and subjective probability"), my views differ from Callen's in a fundamental sense. Both examples given by Callen could be subjective or objective depending on the *given* condition or on the given relevant knowledge.

I have quoted Callen's paragraph above to show that his argument favoring "disorder" is essentially fallacious. I believe Callen has misapplied probabilistic argument to deem information "subjective" and to advocate in favor of "disorder," which in his view is "objective."

An extraterrestial visitor, who has no information on the recorded gender of newborn infants, would have no idea what the probabilities for a male or female are, and his assignment of probabilities would be totally subjective. On the other hand, given the same information and the same knowledge, including the frequencies of boys and girls, the reliability of all the statistical medical records, his assignment of probabilities will inevitably be *objective*.

It is unfortunate and perhaps even ironic that Callen dismisses "information" as subjective, while at the same time embracing Shannon's definition of information, but renaming it as disorder. By doing that, he actually replaces a well-defined, quantitative and objective quantity with a more subjective concept of disorder. Had Callen not used Shannon's definition of information, the concept of disorder would have remained an undefined, qualitative and highly subjective quantity.

In my view, it does not make any difference if you refer to *information* or to *disorder*, as subjective or objective. What matters is that order and disorder are not well-defined, scientific concepts. On the other hand, information is a well-defined scientific quantity, as much as a point or a line is *scientific* in geometry, or the mass or charge of a particle is *scientific* in physics.

Ilya Prigogine (1997) in his recent book *End of Certainty* quotes Murray-Gell-Mann (1994), saying:

> *"Entropy and information are very closely related. In fact, entropy can be regarded as a measure of ignorance. When it is known only that a system is in a given macrostate, the entropy of the macrostate measures the degree of ignorance the microstate is in by counting the number of bits of additional information needed to specify it, with all the microstates treated as equally probable."*[14]

---

[14]The microstates and macrostates referred to here are what we call *specific* and *dim*-configurations, or states, or events.

I fully agree with this quotation by Gell-Mann, yet Ilya Prigogine, commenting on this very paragraph, writes:

"We believe that these arguments are untenable. They imply that it is our own ignorance, our coarse graining, that leads to the second law."

Untenable? Why?

The reason for these two diametrically contradictory views by two great Nobel prize winners lies in the misunderstanding of the concept of information.

In my opinion, Gell-Mann is not only right in his statement, but he is also careful to say "entropy *can* be regarded as a measure of ignorance... Entropy ... measures the degree of ignorance." He does not say *"our own ignorance,"* as misinterpreted by Prigogine.

Indeed, information, as we have seen in Chapter 2, is a measure that *is there* in the system (or in the game of Chapter 2). Within "information theory," "information" is not a subjective quantity. Gell-Mann uses the term "ignorance" as a synonym of "lack of information." As such, ignorance is also an objective quantity that belongs to the system and it is not the same as *"our own ignorance,"* which might or might not be an objective quantity.

The misinterpretation of the informational-theoretical interpretation of entropy as a subjective concept is quite common. I will quote one more paragraph from Atkins' preface from the book *The Second Law.*[15]

" I have deliberately omitted reference to the relation between information theory and entropy. There is the danger, it seems to me, of giving the impression that entropy requires the existence of some cognizant entity capable of possessing "information" or of being to some

---

[15] Atkins (1984).

degree "ignorant." It is then only a small step to the pre-
sumption that entropy is all in the mind, and hence is an
aspect of the observer."

Atkins' rejection of the informational interpretation of entropy on grounds that this "analogy" might lead to the "presumption that entropy is all in the mind," is ironic. Instead, he uses the terms "disorder" and "disorganized," etc., which in my view are concepts that are far more "in the mind."

The fact is that there is not only an "analogy" between entropy and information; the two concepts can also be made identical.

It should be stressed again that the interpretation of entropy as a measure of information cannot be used to *explain* the Second Law of Thermodynamics. The statement that entropy is an ever-increasing quantity in a spontaneous process (in an isolated system) is not *explained* by saying that this is "nature's way of increasing disorder," or "nature's way of increasing ignorance." All these are possible descriptions of the *thing* that changes in a spontaneous process. As a description, "information" is even more appropriate than the term "entropy" itself in describing the thing that changes.

Before ending this section on entropy and information, I should mention a nagging problem that has hindered the acceptance of the interpretation of entropy as information. We recall that entropy was *defined* as a quantity of heat divided by temperature. As such, it has the units of energy divided by K (i.e., Joules over K or J/K, K being the units of the absolute temperature in Kelvin scale). These two are tangible, measurable and well-defined concepts. How is it that "information," which is a dimensionless quantity,[16] a number that has nothing to do with either energy or temperature, could be associated with entropy,

---

[16]I used here "dimensionless" as unit-less or lack of units.

a quantity that has been *defined* in terms of energy and temperature? I believe that this is a very valid point of concern which deserves some further examination. In fact, even Shannon himself recognized that his measure of information becomes identical with entropy only when it is multiplied by a constant k (now known as the Boltzmann constant), which has the units of energy divided by temperature. This in itself does not help much in proving that the two apparently very different concepts are identical. I believe there is a deeper reason for the difficulty of identifying entropy with information. I will elaborate on this issue on two levels.

First, note that in the process depicted in Fig. (8.1c), the change in entropy does involve some quantity of heat transferred as well as the temperature. But this is only one example of a spontaneous process. Consider the expansion of an ideal gas in Fig. (8.1a) or the mixing of two ideal gases in Fig. (8.1b). In both cases, the entropy increases. However, in both cases, there is no change in energy, no heat transfer, and no involvement of temperature. If you carry out these two processes for ideal gas in an isolated condition, then the entropy change will be fixed, independent of the temperature at which the process has been carried out and obviously no heat transfer from one body to another is involved. These examples are only suggestive that entropy change does not *necessarily* involve units of energy and temperature.

The second point is perhaps on a deeper level. The units of entropy (J/K) are not only unnecessary for entropy, but they *should not* be used to express entropy at all. The involvement of energy and temperature in the original definition of entropy is a historical accident, a relic of the pre-atomistic era of thermodynamics.

Recall that temperature was defined earlier than entropy and earlier than the kinetic theory of heat. Kelvin introduced the absolute scale of temperature in 1854. Maxwell published his paper on the molecular distribution of velocities in 1859. This has led to the *identification* of temperature with the mean kinetic energy of atoms or molecules in the gas.[17] Once the identification of temperature as a measure of the average kinetic energy of the atoms had been confirmed and accepted, there was no reason to keep the old units of K. One should redefine a new absolute temperature, denoting it tentatively as $\overline{T}$, defined by $\overline{T} = kT$. The new temperature $\overline{T}$ would have the units of energy and there should be no need for the Boltzmann constant k.[18] The equation for the entropy would simply be $S = \ln W$, and entropy would be rendered dimensionless![19]

Had the kinetic theory of gases preceded Carnot, Clausius and Kelvin, the change in entropy would still have been defined as energy divided by temperature. But then this ratio would have been dimensionless. This will not only simplify Boltzmann's formula for entropy, but will also facilitate the *identification* of the *thermodynamic* entropy with Shannon's information.

In (1930), G. N. Lewis wrote:

"Gain in entropy always means loss of information and nothing more."

This is an almost prophetic statement made eighteen years before information theory was born. Lewis' statement left no

---

[17]This identity has the form (for atomic particles of mass m) $\frac{3kT}{2} = \frac{m\langle v^2 \rangle}{2}$ where T is the absolute temperature and $\langle v^2 \rangle$, the average of the squared velocity of the atoms, and k, the same k appearing on Boltzmann's tombstone.

[18]In doing so, the relation $3kT/2 = m\langle v^2 \rangle/2$ will become simpler $3\overline{T}/2 = m\langle v^2 \rangle/2$. The gas constant R in the equation of state for ideal gases would be changed into Avogadro number $N_{AV} = 6.022 \times 10^{23}$ and the equation state of one mole of an ideal gas will read: $PV = N_{AV}\overline{T}$, instead of $PV = RT$.

[19]Boltzmann's formula assumes that we know what configurations to count in W. To the best of my knowledge, this equation is not challenged within non-relativistic thermodynamics. In the case of Black-Hole entropy, it is not really known if this relation is valid. I owe this comment to Jacob Bekenstein.

doubt that he considered entropy as *conceptually* identical to information.

Shannon (1948) has shown that entropy is *formally* identical with information. There is a story[20] that John von Neumann advised Claude Shannon to use the term "entropy" when discussing information because:

> *"No one knows what entropy really is, so in a debate you will always have the advantage."*

Thus, without entering into the controversy about the question of the subjectivity or objectivity of information, whatever it is, I believe that entropy can be made *identical*, both conceptually *and* formally, to information. The identification of the two is rendered possible by redefining temperature in terms of units of energy.[21] This would automatically expunge the Boltzmann constant (k) from the vocabulary of physics. It will simplify the Boltzmann formula for entropy, and it will remove the stumbling block that has hindered the acceptance of entropy as information for over a hundred years. It is also time to change not only the units of entropy to make it dimensionless,[22] but the term "entropy" altogether. Entropy, as it is now recognized, does not mean "transformation," or "change," or "turn." It does mean *information*. Why not replace the term that means "nothing" as Cooper noted, and does not even convey the meaning it was meant to convey when selected by Clausius? Why not replace it with a simple, familiar, meaningful, and precisely

---

[20]Tribus, M. and McIrvine, E. C. (1971), *Energy and Information*, Scientific American, **225**, pp. 179–188.

[21]As is effectively done in many fields of Physics.

[22]Note that the entropy would still be an extensive quantity, i.e., it would be proportional to the size of the system.

defined term "information?" This will not only remove much of the mystery associated with the unfamiliar word entropy, but will also ease the acceptance of John Wheeler's view to *"regard the physical world as made of information, with energy and matter as incidentals."*[23]

Before concluding this section, I owe you an explanation of my second reservation regarding Cooper's comment cited on page 190.

I agree that "lost heat" could be better than "entropy." However, both the terms "lost heat," and the more common term "unavailable energy," are applied to $T \Delta S$ (i.e., the product of the temperature with the change in entropy), and not to the change of entropy itself. The frequent association of entropy with "lost heat" or "unavailable energy" is due to the fact that it is the entropy that carries the energy units. However, if one defines temperature in terms of units of energy, then entropy becomes dimensionless. Therefore, when forming the product $\overline{T} \Delta S$, it is the *temperature* that carries the burden of the units of energy. This will facilitate the interpretation of $\overline{T} \Delta S$ (not the change in entropy) as either "lost heat" or "unavailable energy."

I should also add one further comment on nomenclature. Brillouin (1962) has suggested to refer to "information" as "neg-entropy." This amounts to replacing a simple, familiar and informative term with a vague and essentially misleading term. Instead, I would suggest replacing entropy with either "neg-information," "missing information," or "uncertainty."

Finally, it should be said that even when we identify entropy with information, there is one very important difference between the thermodynamic information (entropy) and Shannon's information, which is used in communications or in any other branch

---

[23] Quoted by Jacob Bekenstein (2003).

of science. It is the huge difference in order of magnitudes between the two.[24]

As we have seen, the association between entropy and probability not only removes the mystery, but also reduces the Second Law to mere common sense. Perhaps it is ironic that the atomic view of matter that has led to a full understanding of entropy had initially created a new and apparently deeper mystery. This brings us to the next question.

8.4. Is the Second Law Intimately Associated with the Arrow of Time?

Every day, we see numerous processes apparently occurring in one direction, from the mixing of two gases, to the decaying of a dead plant or animal. We never observe the reverse of these phenomena. It is almost natural to feel that this direction of occurrence of the events is in the right direction, consistent with the direction of time. Here is what Greene writes on this matter[25]:

> "We take for granted that there is a direction in the way things unfold in time. Eggs break, but do not unbreak; candles melt, but they don't unmelt; memories are of the past, never of the future; people age, they don't unage."

However, Greene adds: "*The accepted laws of Physics show no such asymmetry, each direction in time, forward*

---

[24] A binary question gives you one bit (binary-unit) of information. A typical book, contains about one million bits. All the printed material in the world is estimated to contain about 10^{15} bits. In statistical mechanics, we deal with information on the order of 10^{23} and more bits. One can define information in units of cents, or dollars, or euros. If it costs one cent to buy one bit of information, then it would cost one million cents to buy the information contained in a typical book. The information contained in one gram of water, all the money in the world, will not suffice to buy!

[25] Greene (2004) page 13.

*and backward, is treated by the laws without distinction,
and that's the origin of a huge puzzle."*

Indeed it is! For almost a century, physicists were puzzled by the apparent conflict between the Second Law of Thermodynamics and the laws of dynamics.[26] As Brian Greene puts it, *"Not only do known laws (of physics) fail to tell us why we see events unfold in only one order, they also tell us that, in theory, events can fold in the reverse order. The crucial question is Why don't we ever see such things? No one has actually witnessed a splattered egg un-splattering, and if those laws treat splattering and un-splattering equally, why does one event happen while its reverse never does?"*

Ever since Waddington associated the Second Law of Thermodynamics with the arrow of time, scientists have endeavored to reconcile this apparent paradox. The equations of motion are symmetrical with respect to going forward or backward in time. Nothing in the equations of motion suggests the possibility of a change in one direction and forbids a change in the opposite direction. On the other hand, many processes we see every day do proceed in one direction and are never observed to occur in the opposite direction. But is the Second Law really associated with the arrow of time?

The classical answer given to this question is that if you are shown a movie played backwards, you will immediately recognize, even if not told, that the movie is going backwards. You will recognize, for instance, that a splattered egg scattered on the floor, suddenly and spontaneously collects itself into the pieces of the broken egg shell, the broken egg shell then becoming

---

[26]Here, we refer to either the classical (Newtonian) or the quantum mechanical laws of dynamics. These are time-symmetric. There are phenomena involving elementary particles that are not time-reversible. However, no one believes that these are the roots of the second law. I owe this comment to Jacob Bekenstein.

whole again, and the egg flying upward and landing intact on the table. If you see that kind of movie, you will smile and invariably conclude that the movie is going backwards. Why? Because you know that this kind of process cannot proceed in this direction in time.

But what if you actually sit in the kitchen one day, look at a splattered egg scattered on the floor, and suddenly the egg gets back to its unbroken state, and then jumps back on top of the table?

Fantastic as it might sound, your association of the process of the splattering of the egg with the arrow of time is so strong that you will not believe what your eyes see, and you will probably look around to see if someone is playing a trick on you by running the film you are acting in backwards. Or, if you understand the Second Law, you might tell yourself that you are fortunate to observe a *real* process, in the *correct* direction of time, a process *that is extremely rare but not impossible.*

This is exactly the conclusion reached by the physicist in George Gamov's book *Mr. Tompkin's Adventure in Wonderland.*[27] When he saw his glass of whisky, suddenly and spontaneously, boiling in its upper part, with ice cubes forming on the lower part, the professor knew that this process, though extremely rare, *can* actually occur. He might have been puzzled to observe such a rare event, but he did not look for someone playing backwards the "movie" he was acting in. Here is that lovely paragraph from Gamov's book:

> " *The liquid in the glass was covered with violently bursting bubbles, and a thin cloud of steam was rising slowly toward the ceiling. It was particularly odd, however, that the drink was boiling only in a comparatively small area*

---

[27]Gamov (1940, 1999).

around the ice cube. The rest of the drink was still quite cold.

'Think of it!' went on the professor in an awed, trembling voice. 'Here, I was telling you about statistical fluctuations in the law of entropy when we actually see one! By some incredible chance, possibly for the first time since the earth began, the faster molecules have all grouped themselves accidentally on one part of the surface of the water and the water has begun to boil by itself.

In the billions of years to come, we will still, probably, be the only people who ever had the chance to observe this extraordinary phenomenon. He watched the drink, which was now slowly cooling down. 'What a stroke of luck!' he breathed happily."

Our association of the spontaneously occurring events with the arrow of time is, however, a mere illusion. An illusion created by the fact that in our lifetime we have never seen even one process that unfolds in the "opposite" direction. The association of the spontaneous, natural occurrence of processes with the arrow of time is almost always valid – almost, but not absolutely always.

George Gamov, in his delightful book *Mr Tompkins in Wonderland*, attempted to explain the difficult-to-accept results of the theories of relativity and quantum mechanics by narrating the adventures of Mr. Tompkins in a world where one can *actually see* and experience the difficult-to-accept results. He tried to imagine how the world would look if the speed of light was much slower than 300,000,000 meters per second, or conversely, how the world would appear to someone travelling at velocities near to the speed of light. In this world, one could observe phenomena that are almost *never* experienced in the real world.

Similarly, one can imagine a world where Planck's constant (h) is very large and experience all kinds of incredible phenomena such as, for example, your car effortlessly penetrating a wall (tunneling effect), and similar phenomena which are *never* experienced in the real world where we live.

To borrow from Gamov's imagination, we can imagine a world where people will be living for a very long time, many times the age of the universe, say $10^{10^{30}}$ years.[28]

In such a world, when performing the experiment with gas expansion, or with mixing of gases, we should see something like what we have observed in the system of 10 dice. If we start with all particles in one box, we shall first observe expansion and the particles will fill the entire volume of the system. But "once in a while" we will also observe visits to the original state. How often? If we live for an extremely long time, say $10^{10^{30}}$ years, and the gas consists of some 10^{23} particles, then we should observe visits to the original state many times in our lifetime. If you watch a film of the expanding gas, running forward or backward, you will not be able to tell the difference. You will have no sense of some phenomena being more "natural" than others, and there should not be a sense of the "arrow of time" associated with the increase (or occasionally decrease) of entropy. Thus, the fact that we do not observe the unsplattering of an egg or unmixing of two gases is not because there is a conflict between the Second Law of Thermodynamics and the equations of motion or the laws of dynamics. There is no such conflict. If we live "long enough" we shall be able observe

---

[28] Perhaps, we should note here that as far as it is known, there is no law of nature that *limits* the longevity of people or of any living system. There might be however, some fundamental symmetry laws that preclude that. But this could be true also for the speed of light and Planck constant. If that is true, then none of Gamov's imaginations could be realized in any "world" where the speed of light or Planck's constant would have different values.

all these reverse processes! The connection between the arrow of time and the Second Law is not absolute, only "temporary," for a mere few billion years.

It should be added that in the context of the association of the Second Law with the arrow of time, some authors invoke our human experience that distinguishes the past from the future. It is true that we remember events from the past, *never* from the future. We also feel that we can affect or influence events in the future, but *never* events in the past. I fully share these experiences. The only question I have is what have these experiences to do with the Second Law or with any law of physics?

This brings me to the next question.

8.5. Is the Second Law of Thermodynamics a Law of Physics?

Most textbooks on statistical mechanics emphasize that the Second Law is not absolute; there are exceptions. Though extremely rare, entropy can go downwand "once in a while."

Noting this aspect of the Second Law, Greene (2004) writes that the Second Law "is not a law in the *conventional sense*." Like any law of nature, the Second Law was founded on experimental grounds. Its formulation in terms of the increasing entropy encapsulates, in a very succinct way, the common feature of a huge number of observations. In its thermodynamic formulation or, rather, in the non-atomistic formulation, the Second Law does not allow exceptions. Like any other law of physics, it proclaims a law that is absolute, with no exceptions. However, once we have grasped the Second Law from the molecular point of view, we realize that there can be exceptions. Though rare, extremely rare, entropy can go the other way. The Second Law is thus *recognized* as not absolute, hence Greene's comments that it is not a law in the "conventional sense." Greene's statement leaves us with the impression that the Second Law is somewhat "weaker" than the conventional laws of physics. It seems to be "less absolute" than the other laws of physics.

But what *is* a law in the conventional sense? Is Newton's law of inertia absolute? Is the constancy of the speed of light absolute? Can we really claim that any law of physics is absolute? We know that these laws have been observed during a few thousand years in which events have been recorded. We can extrapolate to millions or billions of years by examining geological records or radiations emitted from the time near the Big Bang, but we cannot claim that these laws have *always* been the same, or will *always* be the same in the future, and that no exceptions will be found. All we can say is that within a few millions or billions of years, it is *unlikely* that we shall find exceptions to these laws. In fact, there is neither theoretical nor experimental reason to believe that any law of physics is absolute.

From this point of view, the second law is indeed "*not a law in the conventional sense,*" not in a *weaker* sense, as alluded to by Greene, but in a *stronger* sense.

The fact that we *admit* the existence of exceptions to the Second Law makes it "weaker" than other laws of physics only when the other laws are proclaimed to be valid in an *absolute* sense. However, recognizing the extreme rarity of the exceptions to the Second Law makes it not only stronger but the strongest among all other laws of physics. For any law of physics, one can argue that no exceptions can be expected within at most some 10^{10} years. But exceptions to the Second Law can be expected only once in $10^{10000000000}$ or more years.

Thus, the Second Law when formulated within classical (non-atomistic) thermodynamics is an *absolute* law of physics. It allows no exceptions. When formulated in terms of molecular events, violations are permitted. Though it sounds paradoxical, the relative "weakness" of the atomistic formulation makes the Second Law the strongest among other laws of physics, including the Second Law in its thermodynamic (non-atomist) formulation. Putting it differently, the admitted *non-absoluteness* of the atomistic-Second-Law is in fact more absolute than the proclaimed *absoluteness* of the non-atomistic-Second-Law.[29]

In the context of modern cosmology, people speculate on the gloomy fate of the universe, which ultimately will reach a state of thermal equilibrium or "thermal death."

Perhaps not?!

On the other end of the time scale, it has been speculated that since entropy always increases, the universe must have started in the "beginning" with a lowest value of the entropy.

Perhaps not?!

And besides, the last speculation is in direct "conflict" with the Bible:

"1. In the beginning God created the heaven and the earth.
2. And the earth was unformed, and void." Genesis 1:1

א בְּרֵאשִׁית, בָּרָא אֱלֹ הִים, אֵת הַשָּׁמַיִם, וְאֵת הָאָרֶץ.

ב וְהָאָרֶץ, הָיְתָה ת הוּ וָב הוּ, וְח שֶׁךְ, עַל-פְּנֵי תְהוֹם; וְרוּחַ אֱלֹ הִים, מְרַחֶפֶת עַל-פְּנֵי הַמָּיִם.

The original Hebrew version includes the expression "Tohu Vavohu," instead of "unformed" and "void." The traditional interpretation of "Tohu Vavohu," is total chaos, or total *disorder*, or if you prefer, highest entropy!

Having said these, I would venture a provocative view that the Second Law of Thermodynamics is neither "weaker" nor

---

[29] Although my knowledge of cosmology is minimal, I believe that what I have said in this section is applicable also to the "generalized second law," used in connection with black hole entropy, see Bekenstein (1980).

"stronger" than the other laws of physics. It is simply not a law of physics at all, but rather a statement of pure common sense.

This brings me to the last question.

8.6. Can We Do Away with the Second Law?

If the Second Law of Thermodynamics is nothing but a statement of common sense, do we have to list it and teach it as one of the laws of Physics? Paraphrasing this question, suppose that no one had ever formulated the Second Law of Thermodynamics? Could we, by purely logical induction and common sense *derive* the Second Law? My answer is probably yes, provided we have also discovered the atomic nature of matter and the immense number of indistinguishable particles that constitute each piece of material. I believe that one can go from the bottom up and deduce the Second Law.[30] We can certainly do so for the simple example of expansion of gas or mixing two different gases (as we have done at the end of Chapter 7). If we develop highly sophisticated mathematics, we can also predict the most probable fate of a falling egg.[31] All of these predictions would not rely, however, on the laws of physics but on the laws of probability, i.e., on the laws of common sense.

You can rightly claim that I could make this "prediction" because I have benefited from the findings of Carnot, Clausius, Kelvin, Boltzmann and others. So it is not a great feat to "predict" a result that you know in advance. This is probably true. So I will rephrase the question in a more intriguing

---

[30]Here, I do not mean one can deduce the Second Law by solving the equations of motion of particles, but from the statistical behavior of the system. The first is impractical for a system of 10^{23} particles.

[31]Again, I do not mean to predict the behavior of the falling egg by solving the equations of motion of all the particles constituting the egg. However, knowing all the possible degrees of freedom of all the molecules comprising an egg, we could, in principle, predict the most probable fate of a falling egg.

form. Suppose that all these great scientists, who founded the Second Law, never existed, or that they did exist but never formulated the Second Law. Would science arrive at the Second Law purely through logical reasoning, presuming the currently available knowledge of the atomic nature of matter and all the rest of physics?

The answer to this question might be NO! Not because one could not *derive* the Second Law from the bottom up even if no top-down derivation has ever existed. It is because science will find it unnecessary to formulate a law of physics based on purely logical deduction.

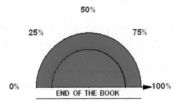

References and Suggested Reading

Atkins, P.W. (1984), *The Second Law*, Scientific American Books, W. H. Freeman and Co., New York.

D' Agostini, G. (2003), *Bayesian Reasoning in Data Analysis, A Critical Introduction.* World Scientific Publ., Singapore.

Barrow, J.D. and Webb, J.K. (2005), *Inconstant Constants*, Scientific American, Vol. 292, 32.

Bekenstein, J.D. (1980), *Black-Hole Thermodynamics, Physics Today*, January 1980 p. 24.

Bekenstein, J.D. (2003), *Information in the Holographic Universe*, Scientific American, Aug. 2003 p. 49.

Ben-Naim, A. (1987), Is Mixing a Thermodynamic Process? *Am. J. of Phys.* 55, 725.

Ben-Naim, A (2006), *Molecular Theory of Liquids*, Oxford University Press, Oxford.

Ben-Naim, A (2007), *Statistical Thermodynamics Based on Information*, World Scientific, in press.

Bent, H.A. (1965), *The Second Law*, Oxford-Univ. Press, New York.

Bennett, D.J. (1998), *Randomness*, Harvard University Press, Cambridge.

Bridgman, P.W. (1941), *The Nature of Thermodynamics*, Harvard Univ. Press.

Brillouin, L. (1962), *Science and Information Theory*, Academic Press.

Broda, E. (1983), *Ludwig Boltzmann. Man. Physicist. Philosopher.* Ox Bow Press, Woodbridge, Connecticut.

Brush, S.G. (1983), *Statistical Physics and the Atomic Theory of Matter, from Boyle and Newton to Landau and Onsager*, Princeton Univ. Press, Princeton.

Callen, H.B. (1985), *Thermodynamics and an Introduction to Thermostatics*, 2nd edition, John Wiley and Sons, US and Canada.

Carnap, R. (1950), *The Logical Foundations of Probability*, The University of Chicago Press, Chicago.

Carnap, R. (1953), *What is Probability?* Scientific American, Vol. **189**, 128–138.

Cercignani, C. (2003), *Ludwig Boltzmann. The Man Who Trusted Atoms*, Oxford Univ. Press, London.

Cooper, L.N. (1968), *An Introduction to the Meaning and Structure of Physics*, Harper and Low, New York.

David, F.N. (1962), *Games, Gods and Gambling, A History of Probability and Statistical Ideas*, Dover Publ., New York.

Denbigh, K.G. and Denbigh, J.S. (1985), *Entropy in Relation to Incomplete Knowledge*, Cambridge Univ. Press, Cambridge.

Falk, R. (1979), *Revision of Probabilities and the Time Axis*, Proceedings of the Third International Conference for the Psychology of Mathematics Education, Warwick, U.K. pp. 64–66.

Fast, J.D. (1962), *Entropy*, The significance of the concept of entropy and its applications in science and technology, Philips Technical Library.

Feller, W. (1950), *An Introduction to Probability Theory and its Application*, John Wiley and Sons, New York.

Feynman R. (1996), *Feynmann Lectures*, Addison Wesley, Reading.

Fowler, R. and Guggenheim, E.A. (1956), *Statistical Thermodynamics*, Cambridge Univ. Press, Cambridge.

Gamov, G. (1940), *Mr. Tompkins in Wonderland*, Cambridge University Press, Cambridge.

Gamov, G. (1947), *One, Two, Three...Infinity, Facts and Speculations of Science*, Dover Publ., New York.

Gamov, G. and Stannard, R. (1999), *The New World of Mr. Tompkins*, Cambridge University Press, Cambridge

Gatlin, L.L. (1972), *Information Theory and the Living System*, Columbia University Press, New York.

Gell-Mann, M. (1994), *The Quark and the Jaguar*, Little Brown, London.

Gnedenko, B.V. (1962), *The Theory of Probability*, Chelsea Publishing Co., New York.

Greene, B. (1999), *The Elegant Universe*, Norton, New York.

Greene, B. (2004), *The Fabric of the Cosmos, Space, Time, and the Texture of Reality*, Alfred A. Knopf.

Greven, A., Keller, G. and Warnecke, G. editors (2003), *Entropy*, Princeton Univ. Press, Princeton.

Guggenheim, E.A. (1949), Research **2**, 450.

Jaynes, E.T., (1957), Information Theory and Statistical Mechanics, *Phys. Rev.* **106**, 620.

Jaynes, E.T. (1965), Gibbs vs. Boltzmann Entropies, *American J. of Physics*, **33**, 391.

Jaynes, E.T. (1983), Papers on Probability Statistics and Statistical Physics, Edited by R.D. Rosenkrantz, D. Reidel Publishing Co., London.

Jaynes, E.T. (2003), *Probability Theory With Applications in Science and Engineering*, ed. G.L. Brethhorst, Cambridge Univ. Press, Cambridge.

Katz, A. (1967), *Principles of Statistical Mechanics, The Information Theory Approach*, W. H. Freeman and Co., San Francisco.

Kerrich, J.E. (1964), *An Experimental Introduction to the Theory of Probability*, Witwatersrand University Press, Johannesburg.

Lebowitz, J.L. (1993), Boltzmann's Entropy and Time's Arrow, *Physics Today*, Sept. 1993, p. 32.

Lewis, G.N. (1930), The Symmetry of Time in Physics, *Science*, **71**, 0569.

Lindley (2001), *Boltzmann's Atom*, The Free Press, New York.

Morowitz, H.J. (1992), *Beginnings of Cellular Life. Metabolism Recapitulates*, Biogenesis, Yale University Press.

Mazo, R.M. (2002), *Brownian Motion, Fluctuations, Dynamics and Applications*, Clarendon Press, Oxford.

Nickerson, R.S. (2004), *Cognition and Chance, The Psychology of Probabilistic Reasoning*, Lawrence Erlbaum Associates, Publishers, London.

Papoulis, A. (1965), *Probability, Random Variables and Stochastic Processes*, McGraw Hill Book Comp. New York.

Penrose, R. (1989), *The Emperor's New Mind*, Oxford Univ. Press, Oxford.

Penrose, R. (1994), *Shadows of the Mind. A Search for the Missing Science of Consciousness*. Oxford Univ. Press, Oxford.

Planck, M. (1945), *Treatise on Thermodynamics*, Dover, New York.

Prigogine, I. (1997), *The End of Certainty, Time, Chaos, and the New Laws of Nature*, The Free Press, New York.

Rigden, J.S. (2005), *Einstein 1905. The Standard of Greatness*, Harvard Univ. Press, Cambridge.

Schrodinger, E. (1945), *What is life?* Cambridge, University Press, Cambridge.

Schrodinger, E. (1952), *Statistical Thermodynamics*, Cambridge U.P., Cambridge.

Shannon, C.E. (1948), The Mathematical Theory of Communication, *Bell System Tech Journal* 27, 379, 623; Shannon, C.E. and Weaver, (1949) W. Univ. of Illinois Press, Urbana.

Tribus, M. and McIrvine, E.C. (1971), Entropy and Information, *Scientific American*, 225, pp. 179–188.

DISORDERED?

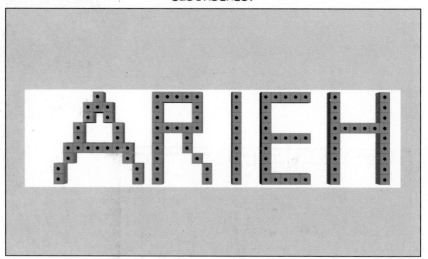

DISORDERED?